Environmental Sampling
for Hazardous Wastes

ACS SYMPOSIUM SERIES **267**

Environmental Sampling for Hazardous Wastes

Glenn E. Schweitzer, EDITOR
U.S. Environmental Protection Agency

John A. Santolucito, EDITOR
U.S. Environmental Protection Agency

Based on a workshop sponsored by
the Committee on Environmental Improvement
of the American Chemical Society,
the U.S. Environmental Protection Agency,
and the University of Nevada—Las Vegas,
Las Vegas, Nevada,
February 1–3, 1984

American Chemical Society, Washington, D.C. 1984

TD
193
.E58
1984

Library of Congress Cataloging-in-Publication Data

Environmental sampling for hazardous wastes.
  (ACS symposium series; ISSN 0097-6156; 267)

  Includes bibliographies and indexes.

  1. Environmental monitoring—Congresses.
2. Hazardous wastes—Congresses.

  I. Schweitzer, Glenn E., 1930–   . II. Santolucito,
John A. III. American Chemical Society. Committee
on Environmental Improvement. IV. United States.
Environmental Protection Agency. V. University of
Nevada, Las Vegas. VI. Series.

TD193.E58   1984   363.7′28   84–20480
ISBN 0–8412–0884–0

FOREWORD

The ACS SYMPOSIUM SERIES was founded in 1974 to provide a medium for publishing symposia quickly in book form. The format of the Series parallels that of the continuing ADVANCES IN CHEMISTRY SERIES except that in order to save time the papers are not typeset but are reproduced as they are submitted by the authors in camera-ready form. Papers are reviewed under the supervision of the Editors with the assistance of the Series Advisory Board and are selected to maintain the integrity of the symposia; however, verbatim reproductions of previously published papers are not accepted. Both reviews and reports of research are acceptable since symposia may embrace both types of presentation.

CONTENTS

PREFACE

THE IMPLEMENTATION of the Resource Conservation and Recovery Act (RCRA) and the Comprehensive Environmental Response, Compensation, and Liability Act (Superfund) has underscored a number of the weaknesses in our capabilities to measure the chemical characteristics of wastes. We are now being called upon to identify and quantify with unprecedented sensitivity hundreds of chemicals found in many types of materials within waste sites, near discharges of hazardous contaminants, and in the surrounding environments. Extrapolations from a limited number of measurements must indicate the general environmental conditions near waste sites. The measurements have to be made faster and cheaper than ever before, with the precision and bias of each measurement fully documented. Thus, the technical challenges facing the monitoring community are substantial.

The progress to date in responding to these challenges has been impressive. Many governmental, industrial, and academic laboratories have become equipped with a new generation of computer-based instruments, and they bear little resemblance to the laboratories of a decade ago. Field monitoring activities are becoming far more sophisticated in design and in implementation. Remote sensing tools guide our sampling efforts, and computer models help interpret our data. But we have only begun to exploit the promise of technology to penetrate the earth, to discriminate among molecular structures, and to allow us to choose those few samples that will adequately represent the whole.

A great deal of operational monitoring activity is underway throughout the nation while the regulatory basis for this activity continues to expand. Many of the monitoring requirements and the associated research needs to support these efforts are clear although new technical issues are constantly emerging as more practical problems unfold. The monitoring responsibility rests with the U.S. Environmental Protection Agency, the states, and the operators of disposal sites. However, the necessary research can only be accomplished through the concerted efforts of a far larger number of organizations.

Perhaps the most neglected aspect in our haste to advance rapidly in assessing and cleaning up waste sites has been the technical basis for the design of environmental sampling programs. The costs of analyzing the large numbers of samples being collected is now causing more careful consideration of the importance of each sample. A second concern relates to the availability of appropriate sampling equipment and techniques. Sample integrity, with particular attention to cross contamination among samples, preservation, and general handling procedures, has at last been recognized as a key to obtaining reliable monitoring data. Since both sampling and analysis comprise the complete chemical measurement system, the quality assurance aspects of sampling programs, and particularly external evaluations and audits, are no less important than the elaborate quality assurance measures that have received so much attention within the analytical laboratories.

In December 1983, the Committee on Environmental Improvement of the American Chemical Society published in *Analytical Chemistry* "Principles of Environmental Analysis." That paper emphasized the importance of sound quality assurance procedures for chemical analyses in the laboratory. Little attention was devoted to sampling concerns. This book addresses some of the important considerations in designing and implementing sampling programs, with particular attention to surface and subsurface sampling for hazardous wastes. Furthermore, it reflects the experiences of federal and state agencies and of academic and industrial organizations and provides a good introduction to the subject. However, considerable additional research and synthesis of experiences are clearly in order given the costs, and more importantly the environmental stakes, involved in obtaining reliable monitoring information.

We hope this book will stimulate greater attention to ensuring that the samples taken to the laboratory or analyzed in the field are indeed the appropriate samples for characterizing contamination problems.

GLENN E. SCHWEITZER
JOHN A. SANTOLUCITO
U.S. Environmental Protection Agency
Las Vegas, Nevada

August 1984

1

Hazardous Waste
Questions and Issues from the Field

H. PATRICIA HYNES

U.S. Environmental Protection Agency, Region 1, Boston, MA 02203

Notwithstanding federal hazardous waste statutes which address the dangers to human health and the environment posed by hazardous wastes, the compliance engineer is presented with unique and demanding challenges. Neither §7003 of the Resource Conservation and Recovery Act (RCRA) nor §106 of the Comprehensive Environmental Response, Cost, and Liability Act (CERCLA) sets forth the levels or quantities of hazardous substances or wastes which constitute imminent and substantial endangerment. Secondly, although there are a few "action levels" developed under the Toxic Substances and Safe Drinking Water Programs which may be adapted for cleanup, neither of the hazardous waste statutes sets forth cleanup levels to be attained for particular contaminants or for generic environmental situations.

A number of site-specific factors must first be evaluated, including (1) the chemical characteristics and amount of hazardous waste, (2) the potential for release to the environment, (3) the sensitivity of the particular environment to the hazardous waste, (4) the proximity of the hazardous waste to humans, and (5) its potential effect on human health. Then the environmental engineer must decide if a field investigation of the site is necessary, whether a feasibility study for remedial action is required, what remedial action is required to mitigate, if not eliminate, the contamination, and finally, what monitoring plan will enable the efficacy of the remedial action to be evaluated.

These decisions have substantial economic impact. They involve the rigorous integration of toxicology and other environmental sciences. Futhermore, they must be defensible in court if challenged.

Environmental scientists and engineers have been working almost four years in the federal hazardous waste program conducting field investigation studies, comparative remedial action studies, emergency cleanups, and to a lesser extent, implementation of long-term remedial action measures. The steep learning curve of these few years has generated a suite of questions and issues common to many hazardous waste sites.

Types of Technical Concerns

With regard to landfills, the following questions arise. Should the contents of a problem landfill be relocated to a more environmentally secure landfill; should chemical and thermal destruction be used when feasible; or should wastes be left in place and preventive measures used to minimize leachate discharge? If the contents of a heterogeneous landfill are unknown, what conditions would be appropriate for the direct sampling of the landfill? What remote sensing technology would be most effective in the detection of buried containers and contaminated ground water plumes and in estimating the depth of each? Also, can remote sensing determine whether the landfill is in direct contact with ground water at any time of year, and the depth to bedrock under the disposal site?

Another difficult area is ground water monitoring. How many wells should be installed; where should they be placed with respect to the site; should they routinely extend to and into bedrock; does an "impermeable" till layer always serve as a vertical boundary for ground water investigation; what length of well screen is optimum? Regarding the final question, as an example of more detailed concerns, a number of factors must be considered. The longer the screen, the better the chance of detecting ground water contamination in an area where contamination exists in a saturated zone if the zone is fairly deep. However, the longer screen may result in greater dilution of contaminants, if they are present in a shallow plume. Low level contamination would, therefore, be more difficult to detect. A shorter screen could miss the zone of contamination depending on the placement of the screen above or below the zone, or if there are dramatic seasonal changes in the water table. Thus, greater accuracy is required in the placement of a short well screen.

Whether to renovate contaminated ground water, eliminate the point source of contamination and leave the ground water to renovate itself, or do both is also a complex decision. For example, earlier manufacturing plants were usually located on or near major rivers or tributaries and often discharged liquid and viscous wastes to unlined lagoons. The subsurface hydraulics often resulted in lagoon discharges to ground water with regional flow patterns toward streams or rivers. If the lagoon is closed for further use and its contents are removed, can the residual contaminated ground water be left to natural cleansing and limited cleanup funds be more usefully spent elsewhere? At a minimum, the decision process must consider (a) present and future uses of the downstream surface water which is the surface expression of the contaminated ground water; (b) present and future uses of any aquifer which under pumping conditions may be infiltrated by the river or the contaminated saturated zone; (c) the mass of contaminants remaining in ground water; (d) the rates of ground water movement and contaminant migration; (e) the combined effect of dilution of the contaminants by ground water and surface water; and (f) any ambient water and drinking water quality criteria which may apply. All of the above must be considered under worst-case conditions.

In a rural area, where leachate from a hazardous waste disposal site has contaminated an aquifer currently used by only a few residential wells, it may appear more cost-effective to replace a water

supply with a pipeline from a nearby municipal system rather than
treat the contaminated ground water. However with this option,
should planning and finance capacity be designed for present or
future use, especially if it is an area projected for high growth?
Further, the quality of the replacement water, if it is treated
surface water, may be significantly lower than that of the original
ground water.

Generic Issues

This paper focuses on issues which are relevant to all hazardous
waste site investigations, remedial actions, and ongoing surveillance
of cleaned sites. Some questions and concerns of field engineers and
scientists are:
- number of samples
- length of sampling period
- frequency of sampling
- indicator parameters and field screening techniques
- methods of quantitative analysis

Two examples, taken from actual case studies, can be given in
which the significance and meaning of the field investigation results
hinged entirely on the design of the environmental sampling programs.
The purpose of the examples is to shed light on those aspects of environmental study which need guidance and additional rigor, especially from the sciences of environmental chemistry and statistics.

The concept of selecting indicator contaminants in environmental sampling appears logical, efficient, and cost-effective. It is
an approach adopted from the RCRA ground water regulations. These
regulations require the owner or operator of a surface impoundment,
landfill, or land treatment facility for hazardous waste to implement
a ground water monitoring program capable of detecting an impact on
the quality of the uppermost aquifer underlying the facility. The
owner or operator must establish initial background concentrations of
parameters regulated under the Safe Drinking Water Act, ground water
quality parameters (e.g., chloride, iron, and phenols), and indicators of ground water contamination (called indicator parameters), and
specifically, pH, Specific Conductance, Total Organic Halogen (TOH),
and Total Organic Carbon (TOC). During the first year of sampling,
for each of the indicator parameters, at least four replicates must be
obtained quarterly and the initial background arithmetic mean and a
variance determined by pooling the replicate measurements. After the
first year, all monitoring wells must be sampled semi-annually. For
each indicator parameter the arithmetic mean and variance, based on
at least four replicate measurements on each sample for each well
monitored, must be calculated and compared with the initial background arithmetic mean. The comparison must consider each of the
wells in the monitoring system and must use the Student's t-test at
the 0.01 level of significance to determine statistically significant
increases (and decreases in the case of pH) from background. If the
initial study reveals elevated levels of any indicator parameter in a
downgradient well, a plan for further definition and investigation of
the contamination must be submitted to the Agency.

The concept of indicator parameters as recommended under RCRA
for industries with a known waste stream with a limited number of

compounds is being proposed for at least two uses by consulting firms:

1. Initial field investigations of old, heterogeneous disposal sites where waste types are insufficiently documented, thus deferring expensive priority pollutant analysis until the fact and location of ground water contamination is established.

2. Design of long-term monitoring for ground water contamination at a waste site following either remedial work or a no-action decision. When ground water contaminants are numerous, parameters are considered less-costly, but "satisfactory," for monitoring trends in ground water quality.

An example of the above is a rural site where for ten years waste oil and solvents from a transformer manufacturing plant were discharged, burned, and buried. This occurred thirty years ago. A recently completed field investigation revealed that highly concentrated PCBs have remained adsorbed to soil in the vicinity of the original disposal and burning area and that eleven volatile organic compounds and one base/neutral compound have migrated as far as 1/2 mile in ground water in the upper aquifer. The wells installed closest to the disposal site contain all of the twelve organics detected in an initial priority pollutant scan and contain the highest concentration of total volatile organics. As the plume attenuates away from the site, the sampling wells contain some but not all of the twelve organic compounds. The study concludes that three compounds, 1,1-dichloroethylene, trans 1,2-dichloroethylene, and trichloroethylene, "are reliable chemical indicators" of seepage from the disposal pit. There is no dicussion of why these three compounds were selected rather than vinyl chloride or other compounds which occur throughout the two major ground water plumes. The selection of the three compounds as "indicators" appears arbitrary and potentially misleading since the chemical and physical properties of all the compounds were not considered.

The responsible party has agreed to remove the original waste and to replace the water supply from private wells with a pipeline from a municipal source, provided they not be obligated to treat the contaminated ground water. Assuming that the ground water is left untreated to renovate itself, there is still reason to have a carefully designed monitoring plan, sufficiently comprehensive and reliable to verify the changes in ground water quality resulting from natural processes. The purpose of a monitoring plan is to provide reliable data to demonstrate with a good degree of certainty whether or not contaminant attenuation results after remedial action. At a minimum, the plan should consider the physical and chemical properties of the compounds in question as well as the soils in the upper aquifer, the ground water flow regime, and statistical techniques to enable meaningful comparisons. For a ground water monitoring plan to be approved by a regulatory agency, it should contain sampling and statistical design features which permit the determination of, for example, whether or not 40 μg/l TCE in Well A in July is a significant increase from 24 μg/l measured in March of the same year. Alternatively, one should know if 35 μg/l benzene in Well A is significantly different from 50 μg/l benzene in Well B if both wells were sampled at the time. Statistical tools are needed to account for the variability of ground water quality, the effect of sampling frequency on the interpretation of data, and the trends in ground water quality.

Another example which poses many questions is a study of PCBs in river sediment and fish and PCB transport under varying flow conditions. The objectives of the field study were:

- o To describe the extent of PCB contamination of fish as a result of former discharges of PCBs into the river by a particular industry.
- o To compare the PCB levels in fish with the Food and Drug Administration (FDA) regulatory level of 5 ppm.
- o To establish a set of initial conditions for measuring impacts of future remedial action on the river.

For comparison between PCB levels in fish and sediment, a river was divided into eight fish and sediment stations. A total of 721 fish (principal species were perch, sunfish, bass, and trout) were collected between 1980 and 1982. Fifty-three percent were used for analysis of PCBs, and those remaining were archived for future use. Filets from each of the four fish species were composited and analyzed. The composites comprised from two to twelve filets, depending on the weight of the fish. The analytical results were reported as a mean concentration in micrograms of PCB per gram of dry weight fish tissue, and consisted of one result per species per station.

Certain limitations of this study are worth reviewing. The FDA standard or "action level" for PCBs in fish sold in interstate commerce was used even though the river is not commercially fished. In any event, a study of contamination in fish which will result in comparisons with a regulatory "action level" for justifying a remedial action must provide results which retain some degree of confidence.

·The sample population for the study (53% of 731 fish) appears at first glance sufficiently large for inferences about the average PCB concentration in the total population. The investigator should decide ahead of time what degree of certainty and interval size are desired. With some estimate of the standard deviation of the population, the sample size required to satisfy the investigation specifications may be determined. However, the iterative process involved in selecting an appropriate sample size was not used in designing this study, nor were the results analyzed and reported with any degree of certainty and interval size. In fact, because of the method of compositing samples the sample population is small—one result per species per station or 8 results per species. In this case, small sample statistical techniques are appropriate.

The design of this fish study centered on sample collection, preservation, preparation, analysis, and QA/QC. There was no discussion of the effect of compositing on the sample population. No description was given of statistical techniques to be applied to the data for reporting results and for comparison with "action levels" and future data. Unfortunately, the omission of a statistical framework during planning of the field study is the rule rather than the exception in hazardous waste investigations.

The hydrogeological and QA/QC aspects of hazardous waste field investigations are fairly well advanced. Yet needed, however, is a systematic approach to the design of field sampling, to the selection of compounds for analysis, and to the methods for interpretation of analytical data.

RECEIVED August 14, 1984

2

Uses of Environmental Testing in Human Health Risk Assessment

CLARK W. HEATH, JR.[1]

Centers for Disease Control, Atlanta, GA 30333

The process for assessing environmental health risks is complex.
Information regarding 1) the environmental agent, 2) the pathways
by which exposure occurs, and 3) the biologic effects observed after
exposure must be assembled and simultaneously evaluated. Incomplete
knowledge or inadequate methodology in any of these three areas can
severely inhibit accurate or useful estimates of health risk.
Environmental testing is a critical element in this process since
it enables the qualitative and quantitative determination of toxic
chemicals in the environment and the definition of environmental
pathways which may lead to human exposure. This paper briefly
reviews the overall process of health risk assessments and the
particular role which environmental testing plays. Recent efforts to
assess environmental health risks in relation to Love Canal illustrate both the usefulness and the limitations of environmental testing in risk assessment.

The Risk Assessment Process

The process of performing risk assessment is outlined in Table I.
Factors contributing to the nature and degree of exposure are
examined first. They include characteristics of the toxic material,
environmental pathways, and mechanisms operating in absorption and
metabolism in the host. Next, the biologic effects resulting from
such exposure are defined in relation to the amount of exposure, to
humans and in animal models. Finally, information about exposure and
biologic effect characteristics is interpreted in relation to predetermined definitions of biologic safety to establish benchmarks for
acceptable exposure risk levels (1).
The toxic chemical(s) of concern must be identified and their
physical and chemical characteristics evaluated. The concentrations
of each of the chemicals must be measured, ideally in both the environment and in the tissues of exposed humans. Depending on the nature
and distribution of toxic material, environmental measurements may be
required in air, water, soil, or food, or in combinations of these
media. The critical limiting factor at this stage of assessment
relates to the degree to which particular chemicals can be identified

[1] Mailing address: Emory University School of Medicine, Atlanta, GA 30322

This chapter not subject to U.S. copyright.
Published 1984, American Chemical Society

Table I.   The Process of Environmental Risk Assessment

I.   ASSESSMENT OF EXPOSURE
   A.   Identification of toxins and measurement of their levels
        or concentrations in the exposure setting.
        1.   In the environment: air, water, soil, food.
             a)   Laboratory technology: quality assurance,
                  precision, sensitivity, accuracy.
             b)   Interaction of chemicals producing potentially
                  toxic byproducts.
        2.   In the host: levels of toxin in serum or tissue.
             a)   Persistence in tissue: excretion, tissue/organ
                  specificity.
             b)   Metabolic alterations: potentially toxic
                  byproducts.
   B.   Environmental pathways for exposure.
        1.   Mechanisms for transmitting toxin to host: ingestion,
             inhalation, dermal contact, physical and biologic
             vectors.
        2.   Degree and mode of exposure: contact through human
             activities.
        3.   Degree and mode of absorption into host: Effective
             dose at tissue/cell level depends on nature of toxin,
             route of exposure, and interaction of target tissue
             with absorbed metabolized toxin.
II. ASSESSMENT OF BIOLOGIC EFFECT
   A.   Human effects: epidemiologic and clinical observations.
        1.   Acute clinical effects: organ system specificity,
             short latency, high dose exposure.
        2.   Delayed (chronic) clinical effects: long and variable
             latency, lower dose exposure.
        3.   Subclinical effects: clinical laboratory test alter-
             ation (liver function, nerve conduction velocity),
             mutagenicity testing, cytogenetic testing. Requires
             estimation of eventual likelihood of clinical disease
             predicted by subclinical abnormalities.
   B.   Non-human effects: experimental observations in toxicologic
        testing in animals or in bacterial or cell culture test
        systems.
   C.   Low dose effects: usually not measurable directly in human
        or animal observations. Need to extrapolate observed high
        dose effects to low or zero dose range by theoretical dose-
        response models.
III. SELECTION OF SAFETY STANDARDS
     Choice of criteria for defining a "safe" level of toxin in the
     environment based on animal and human observations.
     a)   Potential carcinogenic effects: 1/1,000,000 lifetime
          cumulative risk.
     b)   Non-carcinogenic effects: Highest level of toxin at which
          no effect is observed (NOEL), lowered by safety margin
          of 100 to 1000 fold to allow for interspecies biologic
          variation.

---

and measured in each of the media. Testing methodology must be capable of yielding reproducible results of known and acceptable precision and sensitivity. This requirement is especially important when testing is undertaken in more than one laboratory. Although analytic standards and reference materials exist for a wide range of individual chemicals in different environmental media, interactions among multiple chemicals coexisting in the environment pose difficulties for many testing procedures.

Measurement of exposure can be made by determining levels of toxic chemicals in human serum or tissue if the chemicals of concern persist in tissue or if the exposure is recent. For most situations, neither of these conditions is met. As a result, most assessments of exposure depend primarily on chemical measurements in environmental media coupled with semi-quantitative assessments of environmental pathways. However, when measurements in human tissue are possible, valuable exposure information can be obtained, subject to the same limitations cited above for environmental measurement methodology. Interpretation of tissue concentration data is dependent on knowledge of the absorption, excretion, metabolism, and tissue specificity characteristics for the chemical under study. The toxic hazard posed by a particular chemical will depend critically upon the concentration achieved at particular target organ sites. This, in turn, depends upon rates of absorption, transport, and metabolic alteration. Metabolic alterations can involve either partial inactivation of toxic material or conversion to chemicals with increased or differing toxic properties.

Toxic chemicals can be transported with differing levels of efficiency to the target host depending upon the transport pathways. Exposure may occur directly by ingestion, inhalation, or dermal contact or through some form of intermediate vector such as insects or clothing contamination. The relative contribution of different pathways must be assessed by examining the nature of human activities which may be expected in particular exposure settings. This evaluation will identify both the situations for which the greatest exposure may be anticipated (young children ingesting soil while at play, for instance) and the safety standards that will eventually be needed. Again, the actual concentration of toxic chemical in the host cell depends partly on assessments of host-environment contact and partly on knowledge of absorption and metabolism of the particular chemicals.

Biologic Effect. Ideally, risk assessment is based on quantitative knowledge of biologic effects in humans. Unfortunately, such direct human information does not exist for most toxic chemicals. Therefore, prediction of human effects usually depends upon extrapolating the results of experimentally exposed laboratory animals (usually rodents). Such extrapolations must be performed not only between species, but between observed high dose effects and predicted low dose effects. Most animal toxicologic testing and virtually all observed human health effects involve relatively high dosages. Since safety standards are commonly aimed at preventing the potential effects of low dose exposure (especially cancer), low dose extrapolations from existing high dose data are a critical phase in risk assessment. Statistical models for predicting low dose effects exist. Some are based on the assumption that a no-threshold linear relationship exists between dose and biologic response. Others encompass various concepts of threshold effect and curved dose response relationships which predict differing degrees of dose response depending upon the influence of host tissue repair or excretion mechanisms. The eventual safety standards developed can vary quite widely according to the theoretical model selected in a particular risk assessment.

Biologic effects should be assessed for both clinical and subclinical changes. Either can be acute or delayed, with delayed effects often associated with lower exposures. Clinical illness occurs infrequently following chemical exposures, especially at moderate or low dose levels. To increase the probability of detecting clinical endpoints at lower dose levels in experimental animals or human epidemiologic studies, it is desirable to maximize the size of populations examined. Sample size, therefore, rapidly becomes a limiting factor for making clinical observations. When population size is limited, it becomes necessary either to measure subclinical effects which may occur with greater frequency than clinical effects at given dose levels (liver function abnormalities, cytogenetic changes) or to accept theoretical extrapolations downward from high dose clinical effects. Increasingly, risk assessment efforts have begun to focus more on subclinical effects, both in humans and in test animals. Although this trend holds promise for greater testing sensitivity, it will require improved understanding of the relationship between subclinical endpoints and eventual clinical illness. Such subclinical-clinical extrapolation is of critical importance for risk assessment. It is not at all clear at present, for example, that cytogenetic changes observed in exposed populations necessarily fore-shadow later increases in incidence of cancer or genetic disease.

Acceptable Risk. Once information is assembled concerning the characteristics of exposure and biologic effects, that information must be interpreted in terms of human safety standards. That interpretation requires that one establish a set of criteria representing acceptably safe conditions for human existence, bearing in mind that zero concentrations of environmental chemicals are unrealistic.

This process of standard-setting is by nature qualitative and somewhat arbitrary. Nevertheless, certain conventions have evolved for setting safety level targets. In case of carcinogenic or potentially carcinogenic substances, with the assumption that no dose threshold exists for cancer risk, that target has conventionally been set as a lifetime increased risk of one case of cancer in a population of one million persons. For chemicals presumed to be non-carcinogenic, acceptable risk has been conventionally set at the highest dose level at which no observed biologic effect is observed in experimental animals (NOEL or "no observed effects level"). The latter standard is then adjusted downward by applying a safety factor of 100 to 1000 fold to make allowance for uncertainties of extrapolations for differences between species and dosages.

Environmental Testing

Once an acceptable "safe" level of chemicals has been determined for particular environmental media based on existing toxicologic and epidemiologic data and on appropriate safety model criteria, accurate, reproducible, and economically feasible programs for measurement of environmental chemicals must be implemented. Since most toxic chemical exposure situations involve multiple chemicals, the task is far from simple. Aside from the economic feasibilty of testing programs which often involve large numbers of samples, two other considerations are of central concern. These are 1) sampling design and framework, and 2) technical laboratory procedures.

With respect to sampling, sufficient numbers of environmental samples should be obtained to permit reliable statistical and biologic interpretation of results. At the same time, the samples collected should be from environmental locations where human exposure is most likely to occur (or did occur, if questions of past exposures require assessment). They should also be targeted for those environmental media which can be expected to have the greatest potential for human exposure and absorption. Finally, the samples must be obtained and preserved so that the chemicals which pose the greatest threat for human health in terms of toxicity and tissue persistence can be accurately measured.

Laboratory quality assurance procedures must be built into the sampling plan so that reproducibility and precision of test results can be clearly demonstrated when testing is complete. This includes defining the precision of the measurement system and sample collection procedures and providing for adequate numbers of repeat or split samples, as well as blindly inserted positive and negative control specimens. If risk assessment is dependent upon assessing potential environmental exposure, overtime arrangements must be made at the start for laboratory consistency for as long as testing is expected to continue.

Because of the complexity and expense involved, even for limited environmental testing programs involving few chemical toxins, specimen collection and laboratory testing should not be hastily undertaken. Careful advance planning is necessary, complete with outside peer review and approval of proposed testing plans. This is especially true for chemicals for which laboratory technology and sampling procedures are not yet fully developed.

Environmental Testing at the Love Canal

When the Love Canal problem came to active public attention, it was necessary to reconstruct the nature and extent of past exposure as well as address current and future human exposure. Multiple chemicals of uncertain amounts and distribution patterns within the Canal site required varied laboratory technologies, multi-media sampling, and large numbers of samples drawn from a wide and diverse neighborhood setting. The problem was of concern to public health agencies at various levels of Government, and several different test programs were undertaken by different organizations, principally the State of New York in the initial phases of remedial work and the U.S. Environmental Protection Agency (EPA) during later phases.

At the start, the focus was on environmental chemical levels at the Canal site itself and in homes immediately adjoining it. An extensive multi-media testing program was conducted over several months in 1978 and 1979 by the New York State Department of Health (2). This work concentrated on the first two rings of homes adjacent to the Canal in its early phases but later was extended to include larger neighborhoods. Community concern was focused especially to the east, where Canal chemicals might potentially spread through the remaining traces of pre-existing natural drainage channels in surface soil. Although a long list of Canal chemicals was targeted for analysis, a select number of chemicals received particular attention on the basis of their relatively unique presence in the Canal. For example, chlorobenzene and chlorotoluene had value as marker contaminants. Water, soil, and air were sampled, with special emphasis given to water and air within homes where people might be expected to have had the highest and most sustained exposure. In later phases of testing, chemical levels in storm sewers and streams draining the Love Canal neighborhood also received attention. When remedial drainage construction work began, environmental sampling was also required to guide the location of underground drainage pipes and to monitor worker exposure conditions.

Since simultaneous health effect surveys were conducted in the Love Canal area, environmental test results in the homes adjoining the Canal were examined in an effort to demonstrate the presence or absence of correlations between environmental chemical levels and frequencies of particular health abnormalities. This effort was largely unsuccessful, since the total exposed population proved to be too small for meaningful interpretation of most health endpoints of interest and since difficulties in controlling for subjective reporting of health symptoms made it difficult to interpret health survey results (3,4).

The results of environmental testing in Love Canal homes conducted prior to remedial drainage construction were used as a basis for testing the hypothesis that persistent cytogenetic abnormalities might have resulted from Canal exposure in persons who had been living adjacent to the Canal in 1978 (5). For this study, 12 households with the highest concentrations of marker organic chemicals in basement air in 1978 were selected. Persons who had lived in several of these homes were then studied for frequencies of different forms of chromosomal aberration in comparison with frequencies found in simultaneous testing of matched households elsewhere in the Niagara Falls urban/suburban area. No significant differences in frequencies were seen in this comparison. Whether cytogenetic changes were never induced by chemical exposure in homes near the Canal, or if they had been induced but did not persist, could not be resolved by this study.

Extensive testing was also carried out to assess future hazards for human habitation and residential use of the area. This testing was carried out in an extensive program funded by EPA in 1980-81 after remedial drainage construction work at the Canal was complete (6). All of the survey design and technical laboratory problems described above were encountered in this program. Despite extensive efforts to meet requirements for sample size and distribution, to provide for adequate control sampling away from the Canal area, and to allow for satisfactory sample collection procedures and laboratory testing protocols, time and resource constraints limited the scope of the program. Questions were raised regarding the interpretation of results in the face of sustained public concern and litigation over the entire socio-scientific situation. From the viewpoint of scientific assessment, however, review of EPA test results by the U.S. Department of Health and Human Services concluded that, despite limitation in sample size and questions regarding certain technical laboratory procedures, the data were sufficient to judge the Love Canal residential area safe for human residential use. This risk assessment, using the environmental test data in hand, was based on the absence of chemical levels above the low parts per billion range. The conclusion was reached with the stipulation that storm sewer drainage tracks known to contain excess levels of dioxin and other persistent organic chemicals be cleaned, that the Canal site itself not be used for home sites, and that an adequate program for monitoring chemical contaminant within the Canal site be established and maintained into the indefinite future. These recommendations regarding future habitability have not been adopted pending review of the data on which they were based and consideration of the possible need for further evaluative environmental testing.

Literature Cited

1. "Health Risk Estimates for 2,3,7,8-Tetrachlorodibenzodioxin in Soil," Centers for Disease Control, Morbidity and Mortality, Weekly Report, 1984.
2. Kim, C. S.; Narang, R.; Richards, A. et al. Proc. Environmental Protection Agency National Conference on Management of Uncontrolled Hazardous Waste Sites, 1980, p. 212.
3. Vianna, N. J. Proc. 10th Ann. NY State Dept. of Health Birth Defects Symp., 1980, p. 165.
4. Heath, C. W., Jr. Envir. Health Perspect. 1983, 48, 3-7.
5. Heath, C. W., Jr.; Nadel, M. R.; Zack, M. M., Jr., et al. J. Amer. Med. Assoc. 1984, 251, 1437-40.
6. "Environmental Monitoring at Love Canal," Environmental Protection Agency, EPA 600/4-82-030a, 1982.

RECEIVED August 6, 1984

Assessing Cyanide Contamination from an Aluminum Smelter

RICHARD A. BURKHALTER, THEODORE J. MIX, MERLEY F. McCALL, and
DONALD O. PROVOST

Department of Ecology, State of Washington, Olympia, WA 98504

Cyanide contamination of the Spokane aquifer was discovered by Kaiser Aluminum Company personnel in 1978. The Kaiser aluminum reduction facility is located on 240 acres in Mead, Washington, near the northeastern city limits of Spokane (Figure 1). Potliner, which contains cyanide, has been stored at the plant site since 1942. The facility is in a semi-arid region of the state at the foothills of the Rocky Mountain range. The average annual precipitation is 17.5 inches per year with 70 percent occurring from October through April. The annual average evaporation rate is 14 inches per year with a potential of 25 inches per year (1).

The facility is located over the Spokane aquifer which has been designated a major sole source aquifer (Figure 2). The depth to the aquifer at the plant site is about 160 feet. The soil over the aquifer is sand and gravel with interspersing clay lenses (2,3). The Spokane aquifer is a highly permeable aquifer with ground water movement estimated to be between 41 and 47 feet per day by the U.S. Geological Survey and Corps of Engineers. The northern part of the aquifer discharges by springs into the Little Spokane River. Under base flow conditions the flow of the river more than doubles because of these springs (4).

Process and Facility Description

Aluminum is produced by passing an electrical current through a high temperature electrolyte solution (sodium aluminum fluoride) containing aluminum oxide (Figure 3). Aluminum migrates to the carbon cathode (potliner), and oxygen migrates to pre-baked carbon anodes. Aluminum is usually tapped from the pot once every fourth shift. The oxygen supports the combustion of the carbon anode which must be replaced on a routine basis. The potliner lasts a considerable period of time, usually three years, before it fails and is replaced. Approximately 4,500 tons of potliner are discarded each year when operations are at full production levels.

In the process, high temperature and reducing conditions result in the formation of cyanide which is absorbed into the cathode carbon block at the bottom and sides of the pot. When a pot fails, the

0097-6156/84/0267-0015$06.00/0

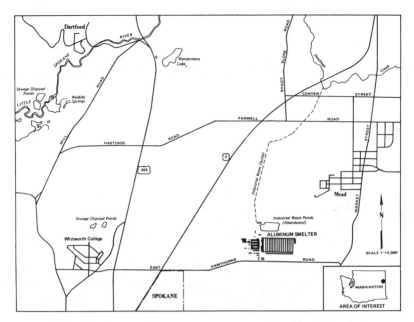

Figure 1.   Vicinity map – Kaiser Aluminum and Chemical
Corporation, Mead Smelter.

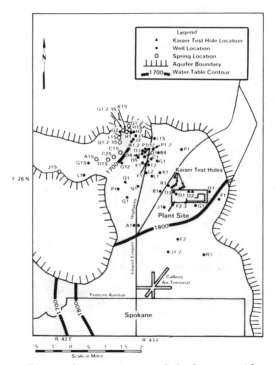

Figure 2.   North arm of Spokane aquifer.

potliner is removed from the steel pot shell. The shell is recon-
ditioned by relining with carbon and insulation before being returned
to the potline.

To recondition the pots, the potliner is dug out and discarded.
Prior to discovery of the Spokane aquifer contamination, the procedure had been to remove the pot to an outdoor concrete slab where the
pot was filled with water and allowed to soak for a few days to frac-
ture and soften the cathode. The contaminated water was presumably
reused for soaking and not discharged to the industrial waste treatment system because of the cyanide content. The pots were jack-
hammered and the potliner dumped on the slab. The potliner was
transfered by a front-end loader to an unprotected pile next to the
slab.

The 8-acre contaminated area is located in the north-central
portion of the plant site (Figure 4). The potliner pile volume is
2.5 million cubic feet (128,000 tons) and contains approximately 0.2
percent cyanide. The Company's industrial wastewater settling basin
(Tharp Lake) was located 200 feet east of the potliner pile. It
removed suspended solids, oil, and grease from several million gallons
of cooling water and storm water runoff each day. The domestic
wastewater treatment plant is also located near the site of the
industrial wastewater treatment facility. The 300,000 gallons per
day of treated domestic wastewater was discharged to a seepage lagoon
located in the same vicinity. The main storm water and industrial
sewer line is located between the potliner pile and the industrial/
domestic treatment systems.

Description of the Problem

Because of forthcoming state and federal regulations and problems
revealed at other plants, the Company decided to drill some test
wells to determine if storage of waste materials at the plant site
had resulted in environmental contamination. Exploratory wells were
drilled by the Company's contractor around the potliner disposal pile
(5). High concentrations of cyanide and fluoride were found in these
wells. As a result of these findings, existing wells outside the
plant boundary were sampled and also found to contain high concentrations of cyanide. The results were reported to the Department of
Ecology in August 1978, and subsequently, additional wells were
tested to determine the extent of the contamination. This sampling
outlined a pathway from the plant to the Little Spokane River, as
shown in Figure 5. The plume is approximately 800 feet wide at the
plant and 1,500 feet wide at the Little Spokane River, a distance of
two and one-half miles northwesterly from the plant site. The ground
water elevation drops 80 feet from the plant site to the Little
Spokane River.

Total cyanide concentrations in ground water samples ranged from
over 300 parts per million (ppm) at the plant site to about 1.5 ppm
at a spring located along the banks of the Little Spokane River.
Wells used for drinking water, irrigation, and livestock purposes
contained total cyanide concentrations as high as 23 ppm.

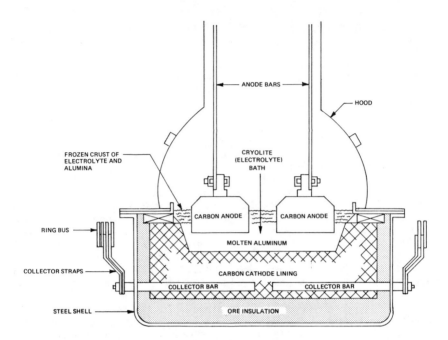

Figure 3. Aluminum reduction cell.

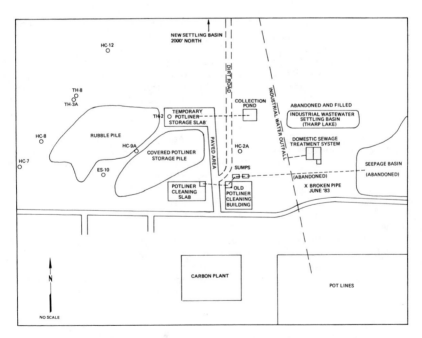

Figure 4. Area of contamination, Kaiser's Mead Smelter.

Remedial Action and Results, Phase I

The Company requested, and immediately received, permission to di-
vert the treated sanitary wastewater from the seepage lagoon to the
industrial treatment system. At about this time, it was also discovered that Kaiser employees were discharging cyanide-laden sump
water directly to the seepage lagoon, and this practice was immediately discontinued. The Company was also ordered to discontinue pot
soaking and instead to dig the pots out in a dry condition. The
Spokane County Health Department ordered the Company to cover the
existing potliner pile. The Company also constructed a temporary
double-lined pad for storage of fresh potliner until they constructed
a storage building. Drainage from the storage slab was discharged to
a new, lined pond for treatment.
    The Company and Spokane County Health Department immediately
contacted all well owners in the vicinity and tested the wells for
cyanide contamination. The Company made bottled water available to
the affected people on a temporary basis until a permanent uncontaminated supply could be obtained. A ground water monitoring program
of selected wells was developed to verify the expected changes resulting from these remedial actions. Additional wells were installed
around the covered pile to support this monitoring program.
    Because of the low rainfall and high evaporation rate, other
water sources which might carry contamination into the ground water
were investigated. The Company was required to check the industrial
water settling basin and all existing storm and sanitary sewers in
the potliner area for leaks.
    A water quality study of the Little Spokane River was conducted
to determine if cyanide discharge by the springs would have any
effect on the aquatic life. Very low concentrations of cyanide were
measured in the Little Spokane River near the contaminated springs,
and no measurable effect on aquatic life was detected.
    The Washington Department of Ecology requested the Company to
conduct a feasibility study of pumping the aquifer and treating the
contaminated ground water. The cost of pumping and treating contaminated ground water was estimated to be over $4 million in capital
costs, with an operating cost of approximately $1 million per year.
If the wells were pumped, the treated water would still contain 200
ppb cyanide and would also need to be disposed of safely.
    Inspection and testing of the storm/industrial and sanitary
sewers in the area indicated the sewers were in good condition and
only minor amounts of leakage were occurring. Nevertheless, minor
repairs were performed.
    As a result of removing the sanitary discharge from the seepage
lagoon, the shallow well (TH-1) located to the west of the seepage
lagoon, showed a decrease in cyanide concentration. The well located
immediately downstream of the potliner pile and potliner work area
did not improve as expected.

Remedial Action and Results, Phase II

The remedial actions summarized here are detailed elsewhere (6-9,10)
By mid-1980 the cyanide levels in the ground water had not changed
significantly, and it was apparent that the corrective action taken

had not improved the contamination problem. The Phase I assumptions and actions were re-examined because more information on the subsurface geology was required to determine the pathways of contamination. One possibility considered was that aquatards (clay lenses) were ponding highly contaminated waters below the pile and slowly releasing them to the main aquifer. Other contamination pathways theorized were:

1. Leaching of cyanide from the uncovered potliner pile into the ground by rainfall and snow melt.
2. Discharge of 1 to 2 percent cyanide-laden water from the pot soaking operation onto the surrounding ground.
3. Pumping of 1 to 2 percent cyanide-laden water to the domestic seepage pond.
4. Seepage of uncontaminated water into the contaminated ground, thus leaching the cyanide into the ground water.

The Company and its consultant decided more monitoring wells were needed to better define the situation, and a new drilling program was initiated.

With the addition of more wells, now totalling over 35, the geological formation under the plant site was better defined. At the potliner site the main aquifer starts about 160 feet below the surface and consists of three zones ("A", "B", and "C"), with "A" zone containing significantly higher concentrations of cyanide than the other two zones. Water flow through the "A" zone was determined to move much slower than the other two zones. Aquatards were found intermittently throughout the formation to the ground water table, and some of these formed saturated zones of highly contaminated water above them. Almost directly under the pile, a ground water mound in "A" zone was discovered. This mound had the effect of shifting the direction of the aquifer flow. Downgradient of the potliner pile, the clay lenses terminate, and the three zones merge into one (Figure 6).

Aquatards could theoretically be redirecting water flows beneath the covered storage pile and leaching out the cyanide from the highly contaminated soil. If so, ground water cyanide concentrations should have decreased after covering the storage pile. Since this did not occur, some other source of cyanide was suspected. The industrial settling basin (Tharp Lake) which originally was believed not to leak significantly was re-checked by diverting the industrial discharge away from the settling basin for 24 hours. The basin was indeed found to be leaking between 50-60 gallons per minute (Figure 7).

The Company immediately began construction of a new settling basin 2,000 feet to the north of the area of concern. Within 60 days the original treatment basin was closed, and within one month following closing, the shallow monitoring wells between it and the pile totally dried up, some within a matter of days. The aquifer flow has shifted back to what is believed to be its normal course, and the ground water mound has dissipated. All other potential water sources in the area have been checked to insure no other carrier is available to leach cyanide from the soil column beneath the pile. All storm drains, sanitary sewers, and pressurized water lines have been re-checked and sealed if necessary. The cyanide levels have been slowly dropping since closing the settling basin (Figure 8).

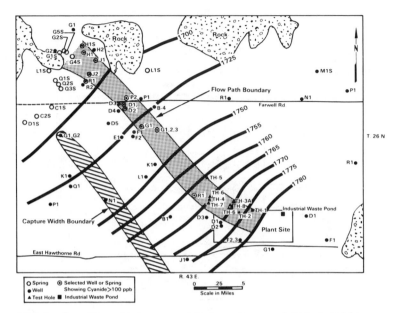

Figure 5. Groundwater cyanide boundary and water table elevations.

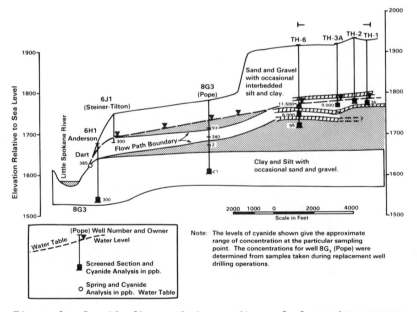

Figure 6. Cyanide flow-path downgradient of plant site sources.

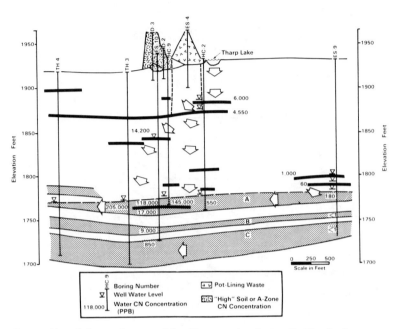

Figure 7. Schematic cyanide flow paths beneath Mead plant.

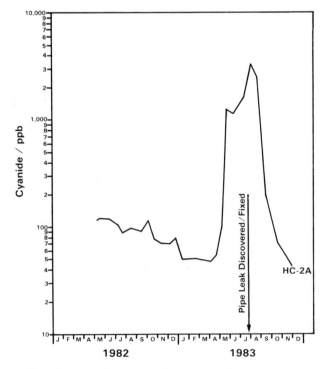

Figure 8. Cyanide levels in five wells/springs, 1981-83.

A leak of 20 gpm in a pressurized water line located in the area to the east of the old potliner cleaning building was observed in late June 1983 and corrected. The results of the leak are vividly shown by an increased cyanide concentration from March 1983 to mid-October 1983 in well HC-2A (Figure 9). This showed the need to carefully control water usage in the contaminated area.

Sampling Problems

Defining the plume of the contamination was straightforward by using existing domestic wells and a few monitoring wells. However, determining the nature of the problem beneath the potliner pile and its vicinity was much more difficult.

Wells installed through the pile were blocked by debris and were generally ineffective. Many wells were drilled around the pile in an attempt to determine the water profile. Preliminary results were deceiving because aquatards and ground water mounding had altered contaminant pathways. The investigators had difficulty determining whether or not leaks were carrying contamination down to the main aquifer.

Analytical Problems

There is a discrepancy between the cyanide criteria for both aquatic and drinking water standards and the current analytical technology. The criteria are stated for free cyanide (which includes hydrocyanic acid and the cyanide ion), but the EPA approved analytical methodology for total cyanide measures the free and combined forms (11). This test probably overestimates the potential toxicity. An alternative method (cyanides amenable to chlorination) measures those cyanide complexes which are readily dissociated, but does not measure the iron cyanide complexes which dissociate in sunlight. This method probably tends to underestimate the potential toxicity. Other methods have been proposed, but similar problems exist (12). The Department of Ecology used the EPA-approved APHA procedure which includes a distillation step for the quantification of total cyanide (13,14). A modification of the procedure which omits the distillation step was used for estimation of free cyanide. Later in the study, the Company used a microdiffusion method for free cyanide (15).

Another potential problem with cyanide analysis is the recommended preservation method. The APHA standard method recommends preservation by adjusting to a pH of 12 using sodium hydroxide. The Department's laboratory has been using this method which is effective for total cyanide but unsatisfactory for free cyanide since the pH adjustment can change the cyanide species present, and thus the final result. There is no adequate preservation method for free cyanide.

In addition to the need for an adequate method for free cyanide and an adequate sample preservation method, a methodology should be developed for the differentiation of species, especially between free (HCN and CN$^-$), metallic complexes, and organic complexes.

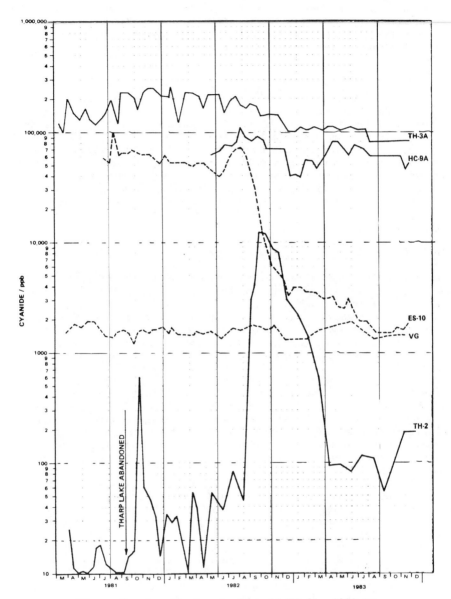

Figure 9.   Cyanide levels in well HC-2A, 1982-83.

Regulatory Considerations

Federal drinking water standards for cyanide have been withdrawn and are not included in the latest publication. The Public Health Service limit for drinking water had been 200 ppb. Whether the limit was expressed as free or total cyanide was in question at the time. The fresh water aquatic cyanide criterion is 3.5 ppb as a 24-hour average, not to exceed 52 ppb at any time.

Potliner waste is exempt from federal RCRA regulations although through State of Washington testing procedures, the potliner at the Kaiser facility was classified as an extremely hazardous waste under state regulations. Because of the ground water contamination of the sole source aquifer, the Environmental Protection Agency has included the Kaiser site in Mead on its Superfund listing.

The Department of Ecology strongly recommended against Superfund status on the grounds that the EPA site evaluation included a population impact based on the number of people who could have been affected in a three-mile radius instead of the population actually affected taking into consideration the directions of ground water movement. Providing the affected residences with a potable water supply by the Company and the impacts of total vs. free cyanide were discussed by EPA but were not used in the impact analysis.

An EPA contractor has prepared a draft remedial action master plan (RAMP) for the Mead site. The contractor recommends further exploration to determine if any undiscovered potliner piles exist and also further geological studies. This recommendation is contingent on cost versus benefit of the action.

Conclusions

This cyanide contamination case study has been an interesting experience because ground water problems are often slow to develop, and cleanup can be even slower.

The major technical problem was the inability to define subsurface geohydrologic conditions with the initial data. Expertise in the area of geohydrology was clearly needed. A lack of specific analytical techniques precluded meaningful environmental and risk assessments. Cleanup efforts were complicated because poltiners are not regulated under RCRA but are regulated under state law. In the middle of the cleanup effort, the site became involved in Superfund activities, and to date this involvement has not been clarified. Project management has become very difficult because of the many players and laws involved. As a result, public confidence has been affected.

Cyanide contamination creates special public information problems, e.g. it is difficult to explain why cyanide is not included in the current drinking water standards but that aquatic organisms are affected at relatively low cyanide concentration. There is confusion on whether fresh water standards are based on free or complexed cyanides. Fortunately, the provision of a permanent drinking water supply to each affected household removed risk assessment as a major issue.

Literature Cited

1. "Normals of Precipitation and Evaporation," State of Washington, U.S. Weather Bureau, February 1961.
2. Cline, D. R. Washington Water Supply Bulletin 1969, No.27.
3. Drost, B. W.; H. R. Seitz. U.S.G.S Open Files Report 1978, 77-829.
4. Burkhalter, Richard A; Cunningham, Richard K; Tracy, Harry B. Technical Report; 1970, No. 70-1.
5. "Effect of Waste Disposal Practices on Ground Water Quality at Kaiser Aluminum and Chemical Corporation: Mead, Washington," Robison and Noble, Inc., April 1978.
6. Dalton, Matthew G. Progress Report, Hart-Crowser and Assoc. March 29, 1983.
7. Dalton, Matthew G. Hart-Crowser and Assoc. Report Sept. 9, 1982.
8. Dalton, Matthew G. Summary Report, Hart-Crowser and Assoc., Dec. 6, 1982.
9. Dalton, Matthew G. Letter, Report; Hart-Crowser and Assoc. April 1, 1981.
10. Dalton, Matthew G. Summary Report, Hart-Crowser and Assoc. Sept. 7, 1983.
11. Ingersoll, D. EPA 600/54-83-054, 1983.
12. "Cyanide: An Overview and Analysis of the Literature on Chemistry, Fate, Toxicity, and Detection in Surface Waters," Ecological Analysts, Inc., 1979.
13. "Standard Methods for the Examination of Water and Wastewater," 15th ed, American Public Health Association: Washington D.C., 1975.
14. "Methods for Chemical Analysis of Water and Wastes," EPA/ 4-79-020, Method 335.2, 1979.
15. Palmer, Thomas A.; Skarset, James Q. Kaiser Aluminum and Chemical Corp"; 1981, No. 81-39.

RECEIVED August 16, 1984

2,3,7,8-Tetrachlorodibenzo-*p*-dioxin Sampling Methods

DANIEL J. HARRIS

U.S. Environmental Protection Agency, Region 7, Kansas City, KS 66115

From October 1982 to October 1983, the Emergency Planning and Response Branch of Region 7 of the United States Environmental Protection Agency and its contractors collected approximately 8,000 environmental samples for analysis of 2, 3, 7, 8-tetrachlorodibenzo-para-dioxin (TCDD). The majority of these samples have been collected and analyzed at an average cost of $700 per sample. This includes per diem, labor, equipment, expendable supplies, transportation, and $400 per analysis by contract laboratories. An evaluation of this data has suggested that field sampling and sample handling methods have a significant impact upon the precision and accuracy of the resulting data which, in turn, impact the cost and feasibility of various remedial options.

Some of the results from sampling at depths to determine the extent of vertical migration of TCDD have been puzzling. Depth samples have been collected in 6- to 12-inch increments down to a maximum depth of 4 feet. For the most part, these samples have been collected along road centerlines where TCDD-contaminated waste liquid oil was sprayed for dust control. Although the surface samples (0- to 6-inch depth) have the highest levels of TCDD, those taken at lower depths have also shown contamination. In some instances, TCDD levels have been higher in the deeper layers than in the overlying ones. With no historical data to offer an explanation for such phenomenon, the data and sampling techniques warrant further examination.

This paper discusses the data resulting from a number of comparative sampling techniques which took place at one well-documented TCDD site.

Study Area in Missouri

The study site, consisting of about 11 contaminated acres belonging to eight property owners, is in a small community midway between Verona and St. Louis, Missouri. Records indicate that on May 20, 1971, a truck driven by an employee of a salvage waste oil company was ticketed for being 950 pounds overweight. The truck was en route from Verona to St. Louis with a load of TCDD-contaminated still

bottoms. Later, it was revealed that to avoid another ticket at a
subsequent weigh station, the owner of the salvage oil company, who
accompanied the driver on the trip, directed the driver to stop at
the owner's farm which is adjacent to the community. It is understood that on the farm the exit valve was opened, and the truck was
driven along a gravel road to dispose of excess waste. The total
quantity of still bottoms off-loaded in this manner over time is not
known. From records of previous loads hauled and truck capacity, it
is reasonable to estimate that the quantity offloaded could not have
exceeded 3,500 gallons. EPA sampling in 1982 confirmed that a county
road in the area was sprayed as well. This finding significantly
expanded the area of suspected surface contamination.

    Although there is no way of knowing with certainty the concen-
tration of TCDD in the off-loaded still bottoms, the waste holding
tank in Verona from which the trucks were filled was sampled in
August 1974 and was found to contain an average of 328 ppm TCDD.

    In total, 550 analyses were conducted from samples taken at this
site. These data indicate that only 5.8 percent of the 10.9 acres
contaminated represented the road surfaces originally sprayed. The
remaining surface contamination probably resulted from dispersion by
wind, vehicular traffic, runoff, etc. The total TCDD sprayed was
probably about 340 grams, with 74 percent still on the areas sprayed.
Mean, volume weighted, TCDD concentrations in the sprayed and dis-
persed areas were 469 and 31 ppb, respectively. Concentrations in
individual composite samples collected from sprayed areas ranged up
to 1,800 ppb. About 90 percent of the TCDD was contained in 13
percent of the soil volume.

Experimental Sampling and Presentation of Data

The sampling took place between August 16 and 24, 1983. Most of the
274 samples collected for shipment to contract laboratories were to
be analyzed for the purpose of more fully delineating contamination
boundaries. Other samples were collected for comparing sample collection and handling techniques.

Areal Variation. One objective of the sampling comparison studies
was to determine the variation in TCDD concentration over a small
area to estimate the error associated with a grab sample concentration. Accordingly, a one-square-yard area was selected adjacent to a
previously sprayed road. The center of the test area was about 6
feet from a sprayed road shoulder. This one-square-yard area was
divided into nine one-square-foot areas. Using clean spoons and
knives, a single scoop was collected from the center of each one-
square-foot area down to a depth of 2 inches. The data are presented
in Figure 1.

    All the analytical data are from the same laboratory; conse-
quently, interlaboratory analytical variation is not a factor. The
intralaboratory variation for that laboratory was 9.1 percent (i.e.,
the relative standard deviation based on repetitive analyses of performance evaluation samples).

Vertical Migration. Historically, surface samples at other TCDD
sites have been taken to depths of 0 to 6 inches using picks and

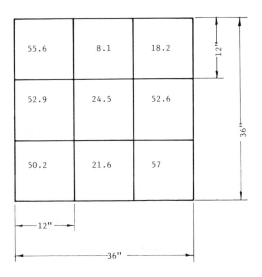

Figure 1. Lateral variation in TCDD concentration, ppb.

shovels. The estimated contaminated soil volumes used in deciding
remedial actions are dependent upon the actual depth of penetration.

To obtain some idea as to the actual vertical penetration of
TCDD in those areas contaminated by dispersion, three sample points
were selected near the farm road which was sprayed in 1971. At each
point a single sample was collected from each of three depths, viz,
at the 0- to 2-, 2- to 4-, and 4- to 6-inch depths, using clean
knives and spoons. The selection of a 2-inch increment was based
upon the estimated removal on a single pass by a qualified operator
of earth-moving machinery. The resulting data are as follows:

	SAMPLE POINT		
	9-B	4-C	10-C
Distance from Road Shoulder (feet)	20	10	15
DEPTHS (inches)	CONCENTRATION (ppb)		
0-2	0.3	5.1	<0.18
2-4	0.51	0.76	<2.20
4-6	<0.16	<0.37	<0.42

TCDD Depth Sampling. Sampling at a depth of 4 feet at other sites has
shown, in some cases, positive TCDD concentrations. Occasionally in
those areas having the lower concentrations, deeper layers have
higher concentrations than the overlying layers. These data have
raised concerns over the actual depth of penetration of the contaminant and the adequacy of sampling methods in obtaining representative
data free of cross-contamination. Previously, depth sampling was
accomplished using augers which were decontaminated between holes or
horizons or both. The auger, although clean initially, must pass
through the relatively high contaminated surface strata before reach-
ing the underlying layers. It is reasonable to expect that the
rotation of the auger will result in smaller soil particles falling
between the hole wall and the auger and consequently being included
in successively deeper samples.

To determine the significance of these concerns, three centerline
sampling segments were selected on the road which was known to have
been sprayed with TCDD-contaminated waste. These segments were
assigned the designations F-1, F-2, and F-3. Segment F-1 was in an
area having soil concentrations in the 1 ppb range (approximate detection limit) at the surface. Segments F-2 and F-3 were in areas of
higher concentrations with F-3 being located in what had previously
been identified as the hotest area on the road. Each centerline
segment, about 15 feet long, was kept as short as possible to reduce
data variability due to longitudinal changes in TCDD concentrations.
In each segment, the following four sampling methods were tried.

1. Averaging Method. A drill rig and auger were used to collect
    columns of soil from depths of 0 to 6, 0 to 12, 0 to 24, 0 to 36,
    and 0 to 48 inches. The holes were drilled about one foot apart
    and the augers decontaminated between holes. Each of the five
    samples were transferred to a clean, stainless-steel pan and
    thoroughly blended prior to splitting into the sample containers.

Consequently, each sample represented an average of the various soils and contaminants in the column.

2. Variable Auger. One hole was drilled, alternately using two different size augers: a 4-inch drill rig auger and a 2-inch manual auger. Starting at the surface, samples were collected in 6-inch increments (eight total), using first the 2-inch auger and then the 4-inch auger to clean out the hole to the top of the next layer. Augers were decontaminated between layers.

3. Shelby Tube. A single hole was drilled and the cores collected at eight successive depths using Shelby Tubes. Each core, after removing about one-half inch from the top and bottom to reduce cross-contamination, was about 5 inches in length. Tubes were decontaminated between horizons.

4. Trench Method. A backhoe was used to dig a trench about 4.5 feet deep and wide enough (30 inches) for field personnel to work. The sampling face of the trench was on the road centerline and positioned so the vertical centerline of the sampling on the trench face would coincide with the location of the Shelby Tube hole previously dug. Starting at the bottom 45-inch level of the trench, clean spoons were used to prepare a fresh face for sampling and to remove any traces of contamination which might have been carried down into the trench by the backhoe bucket. These spoons were then discarded and clean spoons and knives used to collect a sample for analysis. This process was repeated in ascending 6-inch increments up to the 3-inch depth.

Each sample collected in all of the methods was transferred to a clean stainless-steel pan for blending as described in the averaging method.

Figures 2, 3, and 4 present the data on segments F-1, F-2, and F-3, respectively. Because of resistence encountered at F-2 and F-3, it was not possible to complete sampling down to the 4-foot depths by auger and Shelby Tube.

The data presented in Figures 2, 3, and 4 are from five different laboratories. However, the samples from any one method in each segment were sent to the same contractor; thus, interlaboratory analytical variation is not a factor in the data columns, but it may be a factor in the data rows. The estimated interlaboratory and intralaboratory relative standard deviations for TCDD concentrations of 2 to 12 ppb range from 9 to 18 percent and 5 to 13 percent, respectively.

Discussion and Conclusions

The data in Figure 1 show wide variations in the nine single samples collected in a one-square-yard area from a depth of 0 to 2 inches. The low value was 8.1 ppb, and the high value was 57 ppb, with a mean concentration of 37.9 ppb. For practical purposes, a one-square-yard area represents an extremely small point in relation to the size of most sites, and in routine sampling any point within the area would have an equal likelihood of being sampled. Using the nine values, one can arithmetically calculate a number of mean values by taking all possible combinations of the reported values up to 9 at a

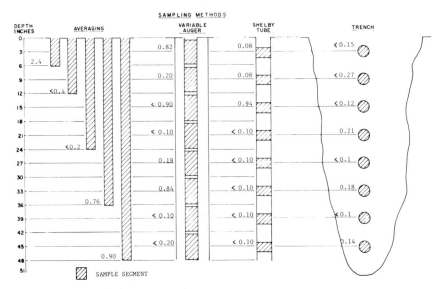

Figure 2. Road centerline TCDD concentrations (ppb) at segment F-1.

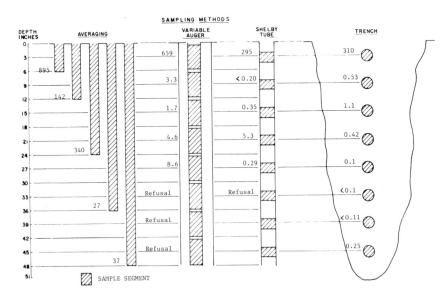

Figure 3. Road centerline TCDD concentrations (ppb) at Segment F-2.

time. This is equivalent to the process of making up 511 composite samples with various numbers of subsamples. The data resulting from this manipulation are as follows:

NUMBERS OF VALUES TAKEN AT ONE TIME	NUMBER OF COMBINATIONS	MINIMUM MEAN ppb	MAXIMUM MEAN ppb	STD DEVIATION ±ppb	STD DEVIATION Relative %
1	9	8.1	57	19.3	51
2	36	14.8	56.3	12.2	32
3	84	16.0	55.2	9.2	24
4	126	18.1	54.6	7.2	19
5	126	24.5	53.7	5.8	15
6	84	29.2	48.8	4.6	12
7	36	32.3	44.9	3.5	9
8	9	35.6	41.6	2.4	6
9	1	37.9	37.9	0	0

The preceding tabulated data are also presented graphically in Figure 5. These data suggest that multiple samples are important in estimating the precision of analytical data. The data also indicate the difficulty in identifying boundaries of TCDD contamination to some designated concentration or advisory level. This variability must be considered in the design of monitoring strategies that provide data for decision-making concerning remedial actions.

The large areal variation in environmental TCDD concentrations is confirmed by the analytical data resulting from using the four different methods to obtain samples at depth. An examination of Figure 3 which presents road centerline data from two 15-foot segments shows variations in the upper six inches ranging, from 295 to 895 and from 2.1 to 1020.

The data resulting from sampling at the 0- to 2-, 2- to 4-, and 4- to 6-inch depths, although not conclusive, suggest that contamination by dispersion is a surface phenomenon and that little vertical migration of TCDD takes place. Unfortunately, none of the three points selected for this sampling proved to be in areas of high contamination. Additional sampling of this nature is recommended.

An examination of Figures 3 and 4 shows clearly that of the four depth sampling mehtods, trenching results in the "cleanest" sampling with the least amount of "drag-in" of contamination to successively deeper layers.

The averaging method can be eliminated as a depth sampling alternative. Using the data from this method, one can calculate theoretical concentrations at various depths, i.e., 6-12 inches, 12-24 inches, etc. It can be shown that in some instances negative concentrations will result. The variable auger method resulted in detectable concentrations of TCDD down to the 24 to 30 inch horizon. The Shelby Tube sampling on segment F-2 (Figure 3) indicated 295 ppb at the surface, below detection at the 9-inch depth, 0.35 ppb at the 15-inch depth, and 5.3 ppb at the 21-inch depth. Looking at the data from the trench method for segments F-2 and F-3 shows TCDD concentrations

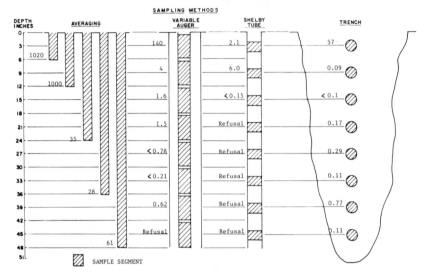

Figure 4.   Road centerline TCDD concentrations (ppb) at Segment F-3.

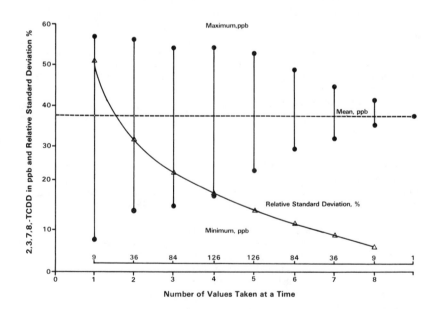

Figure 5.   Minimum and maximum mean TCDD concentrations and relative standard deviations.

decreasing from 310 to 0.53 and 57 to 0.09 ppb, respectively. The detectable trace concentrations are believed to result from having the sampling pan in the trench where fine particles may have fallen into the pan during sampling, and from inherent limitations of the field sampling methods.

The principal conclusions are:

1. Concentrations of TCDD can be expected to vary widely within relatively small areas.
2. Multi-aliquot soil samples are necessary to obtain representative data and to reduce analytical resource expenditures.
3. Depth sampling by trenching provides the "cleanest" samples, i.e., least cross-contamination.
4. TCDD on soil surfaces receiving direct contamination can be expected to be confined to the upper 6 inches of soil.
5. In those areas contaminated by dispersion, TCDD is within the upper 2 inches of soil.
6. Additional sampling is needed to fully characterize the vertical migration of TCDD in those areas contaminated by dispersion.

RECEIVED August 14, 1984

Field Measurement of Polychlorinated Biphenyls in Soil and Sediment Using a Portable Gas Chromatograph

THOMAS M. SPITTLER

U.S. Environmental Protection Agency, Region 1, Lexington, MA 02173

With the recent increase in activity at hazardous waste sites where cleanup and remedial action are underway, there has emerged a need for rapid analytical methods for assessing contamination in water, sediment, and soil. Of special interest, because of widespread use and disposal, is the group of materials known as PCB's (polychlorinated biphenyls).

Equipment and Methodology

The EPA Region 1 Laboratory at Lexington, Massachusetts, has developed a rapid field method for measuring the presence of PCB's in soil and sediments. The analytical technique is GC/EC (gas chromatography using the linearized electron capture detector). Figure 1 shows the portable gas chromatograph used for field work. The column is held in an insulated isothermal chamber. If preheated using AC power, the oven can maintain 200°C for 8 hours on battery operation. The GC has three on-board lecture bottles of gas. These have been manifolded so that 5 cubic feet of 5% methane in argon can be taken into the field. At 30 cc/min, this allows for about one week of operation. The batteries can be recharged overnight for eight hours of use. For PCB measurements a 4' x 1/8" stainless steel column packed with 3% SE-30 on 80/100 mesh chromosorb W-HP is used.

The column is held at 207°C, and most injections are made using 1-3 µl of a hexane extract of soil or a sediment reference material sample. Flow is held at 30 cc/min. These conditions permit elution of the six major peaks of Arochlor 1254 (Figure 2) in about 8 minutes. Field samples are prepared for analysis by weighing out 400 mg of soil into a 2 cc septum vial. When less accuracy is needed, no balance is utilized, and sample size is estimated by volume. To the soil are added 100 µg/l of water, 40 µl of methanol, and 500 µl of technical grade hexane. The sample is agitated for about 20 seconds by hard shaking or by holding the vial to the tip of a vibrating engraver. Finally, a dilution is made if a high PCB level is anticipated. If not, the top layer in the vial is sampled with a 10 µl syringe, and a 1-3 µl sample is injected into the GC.

Figure 1. AID portable gas chromatograph.

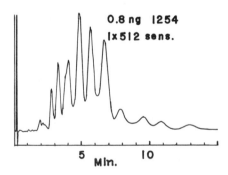

Figure 2. Field chromatogram of Arochlor 1254 extracted from soil.

Results. Various solvent mixtures were tested for extraction efficiency. The test sample was a bone-dry sediment reference material
containing 24.6 ppm of Arochlor 1242. This reference material is a
real sediment from New Bedford Harbor which was homogenized and carefully assayed for PCB's by the Cincinnati EPA facility. Figure 3
shows recovery of 1242 using (1) hexane alone, (2) hexane and water
(1:1), (3) hexane, water, and ethyl ether, (4) ethyl ether and water,
(5) ethyl ether, water, and methanol, (6) methanol and hexane (1:1),
and (7) water, methanol, and hexane (1:4:5). This last combination
appears to give the best recovery. When added in this order to a dry
sample, the effect of the water is to wet the sample, thus permitting
extraction by methanol. The extracted PCB is partitioned almost
exclusively into the hexane from the aqueous methanol. Final recovery is calculated from initial weight and hexane volume.

Quality Control. Figure 4 shows recovery of 1242 from the sediment
reference material in comparison with a hexane standard of pure 1242.
Although other peaks are present, it is evident from the three starred
peaks that recovery of the 1242 is almost complete. Results of
subsequent tests by various chemists at this Laboratory and on the
field investigation team range from 80% to 105% recovery of 1242. In
Figure 5, several standard Arochlors are shown for purposes of pattern
recognition. Figure 6 shows the sensitivity of the method. This is
an injection of 43 pg of chlordane at a sensitivity setting of 1 x
32. Baseline noise is still low enough that 5-10 pg could be readily
determined. This also illustrates the usefulness of a field method
for measuring any other chlorinated compounds of interest.

Precision. The choice of 400 mg of soil is arbitrary. It was chosen
in order to keep the entire cleanup step within a 2 cc vial. A test
of replication was done on one field sample contaminated with about
20,000 ppm of 1254. Three samples of 50, 54, and 54 mg were weighed
into separate vials, extracted, and then diluted 1:1000 into hexane
in a separate vial. The three chromatograms are shown in Figure 7.
Two peaks were quantified to demonstrate how reproducible a measurement can be, even in a field sample.

Field Experience. On the first day of field use, 40 soil and 10 QC
samples were analyzed in six hours. This included lunch and 40
minutes down time when the field generator ran out of gasoline. Most
runs were completed in less than nine minutes, and many very low
level samples had the run aborted after about four minutes when it
was evident that the second major 1254 peak was almost totally absent.
Concentrations were calculated from periodic standard runs, and PCB
levels ranged from less than .2 ppm to 24,000 ppm of 1254.

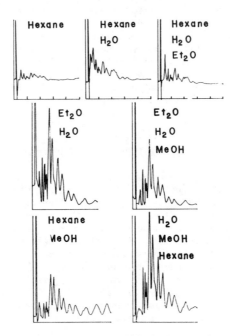

Figure 3. Recovery of Arochlor 1242 from a sediment reference material using different solvents.

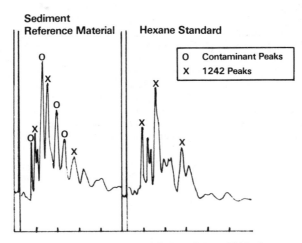

Figure 4. Recovery efficiency of Arochlor 1242 from sediment reference material.

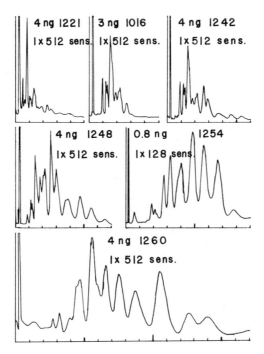

Figure 5.   Reference chromatograms of Arochlor 1221, 1016, 1242, 1248, 1254, and 1260.

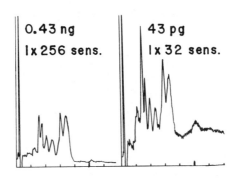

Figure 6.   Sensitivity of field method to Chlordane.

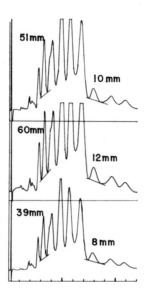

Figure 7. Replication of field measurements of contaminated
soil.

Conclusions

Field measurements provide savings in sample handling and analysis
time, and they eliminate costly delays when re-sampling is required.
In addition, they permit important real-time decisions by the on-scene
cleanup coordinator regarding removal of sufficient contaminated soil
to effect the desired cleanup while avoiding the removal of low-level
contamination beyond that required.

RECEIVED August 14, 1984

Using Geostatistics in Assessing Lead Contamination Near Smelters

GEORGE T. FLATMAN

U.S. Environmental Protection Agency, Environmental Monitoring Systems Laboratory, Las Vegas, NV 89114

The growing number and complexity of toxic chemicals and hazardous waste sites call for more efficient sampling designs and more precise data analysis. Geostatistics is a promising tool for meeting these needs. Four recognized geostatistical texts are cited for general reference (1-4). This paper presents the logic sequence of geostatistical analysis and its application in the Dallas Lead Study (5).

Geostatistical Logic and Tools

The logic of geostatistics is based upon variables being dimensionally correlated. The results of the analysis are presented using three graphic tools: the semi-variogram, the isomap of pollution estimates, and the standard deviation isomap. These tools provide the basis for four advantages that geostatistical sampling design and data analysis can bring to monitoring statistics, namely,

1. A comprehensive theory encompassing sampling design through final error analysis that uses the spatial or temporal correlation of environmental samples to optimize sampling and analysis.
2. A definition of "representativeness" in space or time for input samples and output estimates.
3. An estimate of the value of a sample at any sample site.
4. An estimate of standard deviation at any sample site.

Regional Variable. In pollution monitoring, because of the physical laws governing the source, transport, and fate of the pollutant, samples taken close together have pollutant concentrations closer in magnitude than samples taken farther apart, i.e., each sample is correlated with nearby samples in time or space. Mathematically, this type of sample is described as a regional variable rather than a random variable. A regional variable is not independent and therefore not amenable to classical statistics which is based upon independent random variables, their means, variances, and confidence intervals. Geostatistics is the branch of statistics that describes how to sample and analyze regional variables. Whereas the representativeness of a sample of a random variable is often discussed but seldom rigorously defined, the representativeness of a regional variable is defined in terms of its range of correlation or zone of influence.

A sample is representative of a neighborhood measured by the range of correlation. For example, a soil sample could represent a circular area in the field centered at the sample site with a radius less than or equal to the zone of influence. This has always been intuitively obvious to the environmental scientist but now can be described statistically. The zone of influence is defined by the theoretical semi-variogram and is easily estimated from an empirical semi-variogram.

Semi-Variogram and Range of Correlation or Zone of Influence. The semi-variogram of a regional variable is a mathematical equation and is represented as a graph of the variance of differences between pairs of sample values as a function of distance between those pairs of sample sites. The semi-variogram is plotted as the difference in distance (h) between pairs of sampling sites on an x-axis and half the variance of the difference in pollutant values between these pairs of sampling observations on the y-axis. Second order stationality is assumed, i.e., the mean and variance of the pollutant are considered constant in the area of interest. A change in the mean such as a gradient or cline is called a trend. However, because trend removal does not change basic sampling design, the location function can be simply represented by the distance between points. The semi-variogram should be modeled directionally (N to S, E to W, NE to SW, or NW to SE) because the range of correlation may differ in length in different directions. Differences between directional semi-variograms indicate anisotropy, i.e., directionally dependent discontinuities in the correlation structure. A directionally correlated pathway of deposition can often be anticipated, e.g., prevailing wind direction.

Functions Used To Describe the Variogram

Several different types of semi-variograms are useful. The spherical model with nugget, the random model, and the spherical model with no nugget are discussed below.

Spherical Model with Nugget. Figure 1 shows the spherical semi-variogram model with nugget. The rising curve indicates that the differences between pairs of points closer together have a lower variance than the differences between pairs of points farther apart. This quantifies the empirical insight that samples taken close together in the field are more alike and have less variance than samples taken farther apart. The curve approaches a horizontal value called the sill. It is estimated by the variance of the samples treated as random variables. The distance along the x-axis to the first intersection of the curve with the sill is the range of correlation or zone of influence labeled A in Figure 1A. In this range the curve is below the sill. This shows that a geostatistical analysis has a lower variance and lower error of estimation than a random variable analysis. The y-axis which represents the variance has two parts: one from zero to the y-intercept of the curve and the second from the y-intercept to the sill.

The part below the intercept, labeled C_0 in Figure 1A, is the random variance that cannot be reduced by sample spacing. It has been given the name "nugget effect" which denotes to a gold miner, for example, that the deposit under study is spatially distributed as randomly scattered nuggets rather than as a structured vein or deposit. The nugget effect might also be a "human nugget effect" (2) caused by errors in sample preparation or laboratory analysis. Quality assurance activities should identify a "human nugget" effect and suggest corrective actions. The variance component from the y-intercept to the sill, labeled C_1 in Figure 1A, is the structured variance that can be reduced by sample spacing. In field applications, a predominant component implies that more closely spaced sampling points will give a large improvement in the precision of the output and may be cost effective. A predominant C_0 component or nugget implies that more closely spaced sampling points will give little improvement in the precision of the output and may not be cost effective. The semi-variogram parameters C_0, C_1, and A contain the information needed to optimize sampling design and data analysis.

The empirical semi-variogram computed from the data of the first smelter area during the Dallas Lead Study (Figure 1B) illustrates a spherical model with a large nugget effect. The ordinates of the dots are the variances of the differences of all pairs of points. The abscissas of the dots are the distances (h's, from zero to 3,048 meters) between the pairs of points. The solid line is the sill estimated by the sample variance. Note the general rising of the first eight points and the tight scattering about the sill of the rest of the points. These dots suggest the shape of the spherical model. The last few rising points should be ignored because they represent too few pairs of data values to be reliable, and their abscissas (h) are beyond the cutoff distance of $L/2$. The first few points indicate the y-intercept and the magnitude of the variance components C_1 and C_0, and the first eight or nine dots indicate the length of the range of influence. The abscissa of the intercept between curve and sill is about 366 meters. Thus, in this example the length of the range of correlation is 366 meters. Because of their dependence on small subsets of the data, these parameters (A, C_0, C_1) are more usefully estimated graphically than by techniques for fitting least square curves. The random variance C_0 and the structured variance C_1 are nearly the same, i.e, half of the variance is structural C_1 and can be reduced by sample spacing, but half of the variance is random C_0 and cannot be reduced by design modification.

Random Model. If the range of influence (A) in Figure 1A becomes shorter and approaches zero, then the semi-variogram becomes a random model as shown in Figure 2A. The structural component of the variance C_1 has become zero along with the range of influence, and C_0, the random component, accounts for the whole variance. All these changes denote that the variable sampled is random rather than regional and illustrate the relationship or transition between the two types of variables. That is a random variable is a regional variable whose range of influence has shrunk to zero, or a regional variable is a random variable which is correlated in time and/or space. The empirical semi-variogram of the data from the unpolluted control area in the Dallas Lead Study (Figure 2B) illustrates the

random model. Again, the dots are of the empirical variance of the
differences of all the pairs of points h distance apart up to 1609 m,
and the solid line is the sill estimated by the sample variance.
Note that the first points do not rise, and all points tightly cluster
around the sill. Again, the last few points are meaningless because
there are not enough pairs of samples and the distance between pairs
exceeds the cutoff distance L/2. In the Dallas Lead Study, the first
study area had a smelter at its center resulting in a plume of particu-
late. The samples from this area had a structured semi-variogram.
The reference area had no smelter or plume, and its samples had an
unstructured or random semi-variogram.

Spherical Model with No Nugget. If the random variance C_0 in Figure 1
becomes smaller and approaches zero, then the semi-variogram would
become a spherical model as shown in Figure 3A. The sill, range of
influence, and structured variance (C_1) are still defined and can be
estimated, but now all of the variance is structural. Theoretically,
all variance can be reduced by sampling design. Very precise output
can be obtained if desired and if resources are unlimited. The empirical semi-variogram of a second smelter area from the Dallas Lead
Study illustrates this model as indicated by the ordinate of the
first point in Figure 3B. The difference between the data from the
first and second areas is not a sampling artifact since both areas
had identical grids and the same number of field QC samples.

Isomaps of Pollution Estimates and Standard Deviations

Geostatistics has very understandable and usable outputs. It inter-
polates between sampled points to make estimates for every block of
the monitored sites so that objective isopleths of pollution can be
drawn. Geostatistics uses an algorithm called kriging which gives a
best (minimum variance) linear unbiased estimate of the pollution
value at a point or average for an area. The kriging algorithm is an
interpolation (weighted average) procedure. The interpolation propor-
tions (averaging weights) are called kriging coefficients. They
place most importance on the nearest neighbors of the point or block
being estimated. The size of the neighborhood is determined by the
range of influence from the semi-variogram or by a specified number of
adjacent sample points. If the point to be estimated is a point that
was sampled, the kriging coefficients give the sample value as the
estimate. This recognizes the authority of the data and is in marked
contrast to least square algorithms. In kriging for a block or
point with distant neighbors, the kriging coefficients approach 1/n
or a mean estimate. This acts as a smoothing routine and increases
accuracy by dampening bias (6).
     Figure 4 shows an isomap of the lead pollution in the first
smelter area of the Dallas Lead Study. The round symbol in the center
represents the smelter. The lines are isopleths of lead in soil in
$\mu g/g$. Note the cluster of closed isopleths encircling the smelter.
The large number of concentric isopleths encircling the smelter shows
a steep gradient or rapid change in a short distance between a low
(200 $\mu g/g$) outside and a high (3,000 $\mu g/g$) inside.
     The same geostatistical algorithm, kriging, gives a standard
deviation for each estimate. Thus, the error of interpolation can be

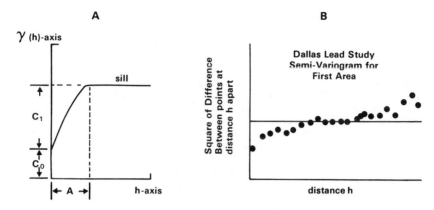

Figure 1A & B. Spherical semi-variogram model with nugget (5).

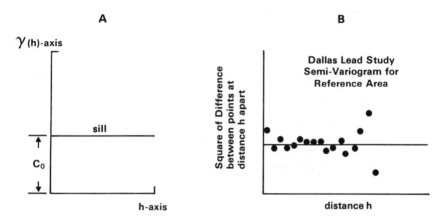

Figure 2A & B. Random semi-variogram model (5).

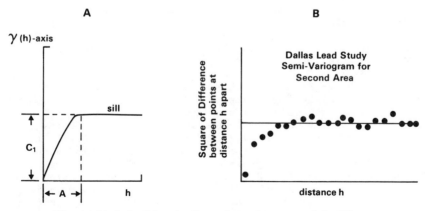

Figure 3A & B. Spherical semi-variogram model (5).

mapped also. This gives isopleths of precision for the concentration
estimates. Figure 5 shows an isomap of interpolation errors. This
is the kriging error map for the first smelter area from the Dallas
Study. The concentric circles show the location of each sampling
site. Where samples were missed as in the lower right corner, the
standard deviation becomes much larger. This isomap confirms the
monitoring scientist's intuition that precision is better closer to a
sample but attaches a numerical value to "better." The pollution
estimates and their kriging errors were calculated for 61 m x 61 m
grid squares approximating the size of city blocks. By using these
two types of geostatistical outputs, a decision maker can identify
areas requiring cleanup, more sampling, or no action. A cleanup
area would be delineated by pollution averages which are above a
chosen critical value and standard errors which are below a chosen
critical value. An area for more investigation would have either
high values for both pollution averages and standard error, or low
pollution averages but high standard error. An area for no action
would have low values for both. In this manner, kriging analysis
extends the usefulness of monitoring data.

Confidence Intervals

The isopleth map of pollution estimates shown in Figure 4 can be
expressed as an isoarea map as in Figure 6. The smelter is located
in the approximate center of the map in the area representing 2500 to
5000 ppm lead. From the smelter the area runs north and then north-
northwest. The wind has a strong and frequent southerly to south-
easterly component. The 2500 to 5000 ppm area is completely enclosed
by the area representing 1000 to 2500 ppm. This level of lead occurs
nowhere else on the map. It is enclosed by the area representing 500
to 1000 ppm lead except on part of its southern edge. This nesting
shows a plume structure enclosing the smelter and is unique in inten-
sity of lead concentration. The plume is a contiguous irregular
polygon, not a string of blocks along the roads. The last three
areas are 250 to 500 ppm, 100 to 250 ppm, and approximately 100 ppm.
They represent urban background levels of lead.
   The isopleth map of errors of estimation for kriging, shown in
Figure 5, is expressed as an isoarea map in Figure 7. The smelter is
located in the approximate center of the map. The kriging error or
standard deviation is multiplicative because the input data were
skewed to the positive side and were transformed by the natural log
before kriging. The lower confidence limit for a multiplicative
standard deviation is found by dividing the estimate by a multiple of
the standard deviation, and the upper confidence limit is found by
multiplying the estimate by a multiple of the standard deviation. A
multiplicative standard deviation of 1.0 would be an exact estimate
or known value. No real-world standard deviation reaches 1.0. The
blank areas have a standard deviation of less than 1.5. If the estimate was 200 ppm, an 80 percent confidence interval for 1.5 would be
$104 < \mu < 384.5$. The open textured areas have a standard deviation
of 1.5 to 1.75. For the estimate of 200 ppm an 80 percent confidence
interval for 1.75 would be $89.2 < \mu < 448.6$. The close textured areas
have a standard deviation of 1.75 to 2.00. Using the standard deviation of 2.0 for the estimate of 200 ppm, an 80 percent confidence

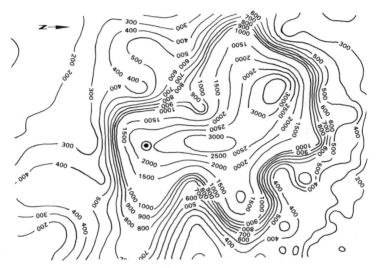

Figure 4.   Isopleth map of kriging estimates of lead concentrations (μg/g) in soil (5).

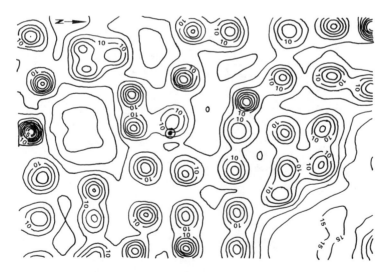

Figure 5.   Ispoleth map of kriging error of estimations of lead concentrations (μg/g) in soil (5).

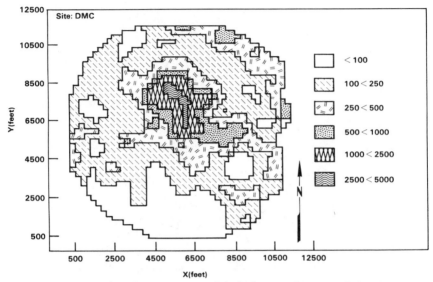

Figure 6. Isoarea map of kriging estimates of lead concentrations (μg/g) in soil (5).

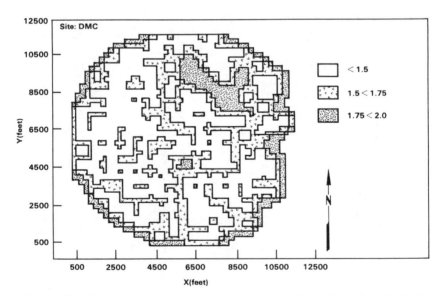

Figure 7. Isoarea map of kriging errors of estimation of lead concentrations (μg/g) in soil (5).

interval would be 78 $\leq \mu \leq$ 512.6. Most researchers become uncomfort-
able with a multiplicative standard deviation greater than 2.0. The
close textured area in the upper right corner is an area where no
samples were taken because of a flooding river. This close texture
also borders the map showing where sampling stopped. The isoarea map
in Figure 7 shows that the field sampling was comprehensive enough to
cover the area of interest and intense enough to ensure a standard
deviation of less than 2.0 except for the river area.

Combining the information from the isoarea map of kriging estimate in Figure 6 and the isoarea map of kriging error or standard
deviations in Figure 7, an 80 percent confidence interval for the
arbitrarily chosen value of 1000 ppm was drawn in Figure 8. Note
that the dark lattice area has about an 80 percent probability of
being above 1000 ppm. Each block on the map (1676 blocks) has had
its own 80 percent confidence interval computed. The block was
assigned to the regularly dotted area if its lower confidence limit
was above 1000 ppm. It was assigned to the light lattice area (below
1000 ppm) if its upper confidence interval was below 1000 ppm. If
1000 ppm fell in its confidence interval, the block was assigned to
the white area (1000 ppm). Note that the area enclosed within the
dotted line is the greater than or equal to a 1000 ppm area from the
kriging estimate map, Figure 4, but a larger area, the white plus the
light lattice areas, is the 80 percent confidence area of greater than
or equal to 1000 ppm. To have 80 percent confidence, more area must
be included in corrective action. This final isoarea map of confidence intervals would be a helpful tool for decision makers.

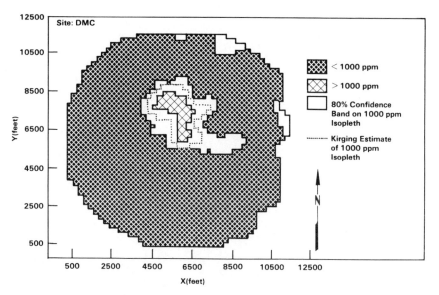

Figure 8.   Isoarea map of 80 percent confidence band on 1000 ppm
lead concentration (5).

Literature Cited

1. Journel, A. G. and Ch. J. Huybregts. "Mining Geostatistics"; Academia Press: London and New York, 1978; pp. 1-597.

2. David, Michel. "Geostatistical Ore Reserve Estimation"; Elsevier Scientific Publishing Company: Amsterdam, Oxford, New York, 1977; pp. 1-364.

3. Davis, John C. "Statistics and Data Analysis in Geology"; John Wiley and Sons Inc.: New York, 1973; pp. 1-545.

4. Clark, Isobel. "Practical Geostatistics"; Applied Science Publishers: London, 1979; p. 130.

5. Brown, K. W.; Beckert, W. F.; Black, S. C.; Flatman, G. T.; Mullins, J. W.; Richitt, E. P.; Simon, S. J. The Dallas Lead Monitoring Study: EMSL-LV Contribution 1983, 2, 10240, 1-265.

6. Efron, B.; Morris, C. Sci. Am. 1977, 119-127.

RECEIVED August 6, 1984

Lead Levels in Blood of Children Around Smelter Sites in Dallas

JOSEPH S. CARRA

U.S. Environmental Protection Agency, Washington, DC 20460

During the fall of 1981, the city of Dallas conducted a voluntary blood-lead screening program involving approximately 12,000 individuals. Analysis of the data from this program suggested that a source of lead existed in the vicinity of the intersection of Singleton Boulevard and Westmoreland Avenue. Expert reviewers concluded that the screening program did not provide definitive conclusions for the following reasons:
- The "sample" was self-selected.
- The study results concerning a smelter as the source were confounded by effects from vehicular traffic.
- Environmental data related to participants in the program were limited.

The reviewers recommended that another study of a valid sample of the preschool population be conducted.

Study Design

In the study reported here, the target populations were limited to preschool children residing in areas of approximately one-mile radius around each of two smelter sites and in a one-half mile radius reference area. The two smelter sites were (1) the RSR site at the intersection of Singleton and Westmoreland, and (2) the Dixie Metal and National Lead smelters (hereinafter referred to as the Dixie site) several miles to the southeast. Each of the two sites was stratified by proximity to the smelter and by proximity to high traffic volume streets. The reference area was chosen based on its similar traffic density to the RSR site and its demographic similarity to the two smelter sites. Housing units were selected in each stratum based on a multistage probability design. The selected housing units were then screened for preschool children. If preschool children were present, questionnaires were administered; blood specimens were collected from eligible children; and soil, paint, and dust specimens were obtained within and around the residences.

The sample size of preschool children was chosen so that specified differences in the percentages of preschool children having blood lead levels greater than or equal to 30 micrograms per deciliter could be measured and compared. Such differences in percentages should occur, given the differences observed between the 0.5 mile radius circle and the 0.5-1 mile ring in the 1981 screening program. Assuming that the data from the 1981 study were reasonable indicators, the chosen sample size should detect gradients in blood lead at the smelter sites and should detect differences between the smelter sites and the reference area.

Table I indicates the sample size and population estimates of preschoolers by study area and stratum. Though all strata for the Dixie site are shown, at the Dixie site the configuration and location of the major traffic artery in relation to the smelter site did not allow analysis of the contribution of vehicular traffic to soil lead. The contribution of the smelter to soil lead levels was possible at the RSR site and the reference site.

Table I. Sample Size and Population Estimates of
Preschoolers by Study Area and Stratum*

	RSR		Dixie		Reference		Combined	
	SS	PE	SS	PE	SS	PE	SS	PE
NH	101	154	30	49	NA	NA	131	203
NL	169	397	76	127	NA	NA	245	523
FH	14	28	29	41	123	224	166	293
FL	216	552	195	369	98	169	509	1090
TOTAL	500	1131	330	586	221	394	1051	2110

Legend: NH - near smelter (0.0 - 0.5 miles) and high traffic area.
         NL - near smelter and low traffic area.
         FH - further from smelter (0.5 - 1.0 miles of reference) and
              high traffic area.
         FL - further from smelter and low traffic area.
         SS - sample size.
         PE - estimate of population size.
*The totals for sample size and population estimates for subsequent tables may not agree exactly with the numbers in this table. This is because information on race for some individuals was not available so that they could not be included in subsequent tables that are broken out by race.

Leaded Paint. The lead content of paint on the interiors of houses was determined by x-ray fluorescence analyzers. The same location was used in each room of each individual house. Fifty percent of the determinations were made on painted walls, and the other fifty percent were made on the trim paint. The precision of these instruments is $\pm$ .2 milligrams per square centimeter. Therefore, values of .7 or greater are considered to represent at least 0.5 mg/cm^2.

Table II shows that houses in the reference area contained significantly more lead paint than houses around either the RSR or Dixie
site, while the RSR area contained more than the Dixie area. Based
on a visual inspection, the paint on the walls and trim areas was
intact and was not peeling, flaking, or otherwise deteriorating.

Table II. Distribution of XRF Readings by Study Area

Study Area	Percentiles						
	5	10	25	50	75	90	95
RSR	.18	.20	.30	.40	.61	.94	1.36
Dixie	.13	.15	.20	.30	.47	.68	.91
Reference	.16	.22	.34	.50	.81	1.44	2.40

Units: mg/cm^2

Traffic Density. Next to the RSR site there is an intersection of
major thoroughfares. One thoroughfare serves approximately 18,000
cars per day, and the other thoroughfare serves approximately 14,000.
One-half mile from the Dixie site there is a major thoroughfare
serving approximately 18,000 vehicles. In the reference area, a
thoroughfare serving approximately 18,000 vehicles intersected the
site. Areas defined as "high traffic density" are within one block
of the roadway. Areas defined as "low traffic density" are more than
one block from the roadway. At the Dixie site the differentiation
between traffic contribution and smelter contribution to the lead in
soil was not possible because of the configuration and location of
the roadway. Because of this, the analysis does not include consideration of traffic density at the Dixie site.

Dust Lead. Table III, IV, and V show the levels of lead in household
dust inside the households in this study. These lead levels were
markedly below what is normally found in household dust. Because
the levels are inexplicably low, the resulting lead levels in the
dust are not considered in the analysis.

Table III. Distribution of Mean Dust Lead Levels by Study Area

Study Area	Percentiles						
	5	10	25	50	75	90	95
RSR	0.00	.02	.04	.07	.11	.18	.27
Dixie	0.00	.00	.00	.03	.07	.13	.20
Reference	0.00	.01	.02	.045	.07	.15	.32

Units: $\mu g/cm^2$

Table IV. Medians of Dust Lead ($\mu g/cm^2$) by
Distance, Traffic, and Study Area

Approximate Distance	Traffic Density	Study Area		
		RSR	Dixie	Reference
Near	High	.080	.000	NA
	Low	.075	.000	NA
	Combined	.080	.000	NA
Far	High	.090	.000	.050
	Low	.060	.040	.040
	Combined	.060	.030	.045
Combined		.070	.030	.045

Table V. 90th Percentile of Dust Lead (ppm) by
Distance, Traffic, and Study Area

Approximate Distance	Traffic Density	Study Area		
		RSR	Dixie	Reference
Near	High	.24	.13	NA
	Low	.21	.20	NA
	Combined	.23	.16	NA
Far	High	.17	.10	.15
	Low	.14	.12	.12
	Combined	.16	.12	.15
Combined		.18	.13	.15

Soil Lead. Tables VI, VII, VIII, and IX depict the soil lead content
at the three sites. The mean, median, and 90th percentile soil lead
levels are higher at the RSR site than corresponding soil lead levels
at the Dixie site which are in turn higher than the corresponding
soil lead levels at the reference site. At the RSR site there were
many more areas of dirt with far less grass covering where children
played than at either the Dixie or the reference sites. At both the
RSR and reference sites, the areas with high traffic density had
significantly higher soil lead content than the low traffic density
areas.

Blood Chemistry Data. All blood specimens were obtained by veni-
puncture. The erythrocyte protoporphyrin (EP) was measured by the
extraction method. Blood lead determinations were done in quadrup-
licate and are presented as an arithmetic mean of the four replicates.

Table VI. Distribution of Mean Soil Lead Levels by Study Area

Study Area	Percentiles						
	5	10	25	50	75	90	95
RSR	78	98	146	288	561	1016	1445
Dixie	35	56	110	191	389	678	995
Reference	72	78	99	135	199	305	463

Units: $\mu g/cm^2$

Table VII. Means of Soil Lead Levels (ppm) by
Distance, Traffic, and Study Area

Approximate Distance	Traffic Density	Study Area		
		RSR	Dixie	Reference
Near	High	1130	NA	NA
	Low	710	NA	NA
	Combined	826	814	NA
Far	High	384	NA	196
	Low	162	NA	148
	Combined	192	193	175
Combined		489	327	175

NA = Not applicable

Table VIII.   Medians of Soil Lead (ppm) by
Distance, Traffic, and Study Area

Approximate Distance	Traffic Density	Study Area		
		RSR	Dixie	Reference
Near	High	768	NA	NA
	Low	521	NA	NA
	Combined	573	553	NA
Far	High	286	NA	153
	Low	150	NA	118
	Combined	164	147	135
Combined		288	191	135

NA = Not applicable

Table IX.   90th Percentile of Soil Lead (ppm) by
Distance, Traffic, and Study Area

Approximate Distance	Traffic Density	Study Area		
		RSR	Dixie	Reference
Near	High	2795	NA	NA
	Low	1252	NA	NA
	Combined	1563	1606	NA
Far	High	457	NA	370
	Low	329	NA	262
	Combined	329	415	305
Combined		1016	678	305

NA = Not applicable

Because the numbers of "other than black" children are small, the analysis is based upon black children. However, the same trends are noted for "other than black" children.

Table X shows the percent of preschoolers having various blood-lead levels at the three sites.

The same information on EP is shown in Table XI.

Table X. Percentage Distribution of Blood-Lead Levels
in Preschool Children

Interval* (µg/dl)	RSR (Percent)	Dixie (Percent)	Reference (Percent)
[0, 5)	0.5	0.7	0.9
[5, 10)	8.0	12.0	17.6
[10, 15)	27.4	33.5	39.3
[15, 20)	34.9	34.1	27.6
[20, 25)	15.6	14.0	11.8
[25, 30)	8.2	3.7	2.8
[30, 35)	4.0	0.8	0.0
[35, 40)	0.8	0.9	0.0
[40, 45)	0.3	0.0	0.0
[45, 50)	0.3	0.0	0.0
[50, +)	0.0	0.3	0.0

*Intervals were calculated using a method equivalent to rounding numbers to integer values before assignment to intervals.

Table XII shows the population estimate, sample size by stratum, and mean blood lead levels among black children in the reference area of 13.2 and 14.6 µg/dl in the low and high traffic density areas, respectively.

There are no children in the reference area who had lead toxicity or whose blood lead level exceeded 29 µg/dl (Table XIII). The term "lead toxicity" is defined here as a child with a blood-lead level ≥ 30 µg/dl and an EP $\geq$ µg/dl. The term "lead toxicity" is not used in a toxicological sense.

Table XIV shows the estimated population of children in the RSR and Dixie sites by race, distance from the smelter sites, and distance from traffic density.

Table XV shows that at the RSR site, the mean blood-lead was found to be 20.1 µg/dl for black children living within 0.5 mile of the smelter (near) as compared to 15.0 for those who live from 0.5 mile to 1 mile (far). Among those who live near the smelter, the mean blood-lead was higher (21.8 µg/dl) for those who also live within 1 block of the major roadways compared to those who live more than 1 block from the roadway (19.1 µg/dl).

Table XVI shows that 17 percent of the black children living near the RSR site, were found to have a blood-lead level ≥ 30 µg/dl in the high traffic density area and 8.3 percent in the low traffic density area. Only 1.6 percent of black children living beyond the 0.5 mile

Table XI. Distribution of EP Levels in Preschool Children

Interval* (g/dl)	RSR (Percent)	Dixie (Percent)	Reference (Percent)
[0, 5)	0.0	0.8	0.0
[5, 10)	0.9	1.2	0.9
[10, 15)	9.9	17.3	15.5
[15, 20)	30.3	29.0	33.9
[20, 25)	22.0	22.0	23.0
[25, 30)	11.3	11.4	10.8
[30, 35)	6.4	5.7	5.9
[35, 40)	5.5	2.8	1.4
[40, 45)	2.9	3.4	3.6
[45, 50)	3.1	0.3	0.5
[51, 55)	1.2	0.6	0.0
[55, 60)	0.8	1.2	1.4
[60, 65)	0.6	0.9	0.4
[65, 70)	1.0	0.6	0.4
[70, 75)	0.7	0.3	0.4
[75, 80)	0.3	0.0	0.0
[80, 85)	0.8	0.3	0.4
[85, 90)	0.2	0.3	0.0
[90, 95)	0.1	0.0	0.0
[95, 100)	1.0	0.6	0.0
[100, +)	1.0	1.2	1.4

*Intervals were calculated using a method equivalent to rounding numbers to integer values before assignment to intervals.

Table XII. Mean Blood Lead Levels by Race in Reference Area

Traffic Density	Estimated Population Size		Sample Size		Mean Blood Lead Level (g/dl)	
	Black	Non-Black	Black	Other	Black	Other
High	186	36	102	20	14.6	13.9
Low	157	12	91	7	13.2	16.8

Table XIII. Distribution of Lead Toxicity and High Blood
Levels by Race at the Reference Area

Traffic Density	Percent with Lead Toxicity		Percent with Blood Lead Levels >30 g/dl	
	Black	Non-Black	Black	Other
High	0	0	0	0
Low	0	0	0	0

Table XIV. Estimated Population by Race,
Distance, and Smelter Site

Approximate Distance (miles)	Traffic Density	Smelter			
		RSR		Dixie	
		Black	Non-Black	Black	Non-Black
0.0 - 0.5	High	134	20	NA	NA
(Near)	Low	376	19	NA	NA
	Total	510	39	99	75
0.5 - 1.0	High	8	18	NA	NA
(Far)	Low	483	64	NA	NA
	Total	491	82	393	11
Total		1001	120	493	86

Table XV. Mean Blood Lead Levels by Race and Distance from RSR Site

Distance	Traffic Density	Sample Size		Mean Blood Lead Level (g/dl)	
		Black	Other	Black	Other
Near	High	88	13	21.8	16.0
	Low	160	8	19.1	11.7
	Combined	248	21	20.1	14.4
Far	High	4	0	14.6	14.9
	Low	189	25	15.3	13.6
	Combined	193	34	15.8	13.6

Table XVI. Distribution of Lead Levels in Blood by
Race and Distance from RSR Site

Approximate Distance (miles)	Traffic Density	Sample Size		Percent with >30 g/dl Lead in Blood	
		Black	Non-Black	Black	Non-Black
0.0 - 0.5	High	88	13	17.0	0.0
	Low	160	8	8.3	0.0
	Combined	248	21	10.5	0.0
0.5 - 1.0	High	4	9	0.0	0.0
	Low	189	25	1.6	0.0
	Combined	193	34	1.6	0.0
Combined		441	55	6.1	0.0

radius from RSR and in the low traffic density areas were found to
have blood-lead levels >30 g/dl.
    Table XVII shows that 5.6 percent of black children living with-
in 0.5 mile of RSR were found to have lead toxicity, and no child
living beyond 0.5 mile of RSR was found to have lead toxicity.
    The mean blood lead level of black children at the Dixie site
was found to be only slightly higher (15.8 g/dl) for those living
within 0.5 mile of the smelter as compared to those living within
0.5 - 1 mile (14.9 g/dl). The three "non-black" children whose
blood-lead levels were 30 g/dl were siblings in a family where a
parent worked at Dixie Metals (Tables XVIII and XIX).

Table XVII. Distribution of Lead Toxicity by
Race and Distance from RSR Site

Approximate Distance (miles)	Traffic Density	Sample Size		Percent with Lead Toxicity	
		Black	Non-Black	Black	Non-Black
0.0 - 0.5	High	88	13	5.7	0.0
	Low	160	8	5.6	0.0
	Combined	248	21	5.6	0.0
0.5 - 1.0	High	4	9	0.0	0.0
	Low	189	25	0.0	0.0
	Combined	193	34	0.0	0.0
Combined		441	55	2.9	0.0

Table XVIII. Mean Blood-Lead Level by Race,
Traffic, and Distance from Dixie Site

Approximate Distance (miles)	Sample Size		Mean Blood Lead Level (g/dl)	
	Black	Other	Black	Other
0.0 - 0.5	60	45	15.8	15.5
0.5 - 1.0	215	6	14.9	17.1

Table XIX. Distribution of Lead Levels in Blood by
Race and Distance from Dixie Site

Approximate Distance (miles)	Sample Size		Percent with >30 g/dl Level in Blood	
	Black	Non-Black	Black	Non-Black
0.0 - 0.5	60	45	1.7	6.6*
0.5 - 1.0	215	6	1.0	0.0
Combined	275	51	1.1	5.7

*This represents three hispanic children in one family where parent had occupational exposure.

Less than 0.5 percent of the black children at the Dixie site were found to have lead toxicity (Table XX). As previously noted, the three non-black children (6.6 percent of non-black children) found to have lead toxicity were siblings in a household where the father had an occupational exposure to lead.

Table XX. Distribution of Lead Toxicity in Blood
by Race and Distance from Dixie Site

Approximate Distance (miles)	Sample Size		Percent with Lead Toxicity	
	Black	Non-Black	Black	Non-Black
0.0 - 0.5	60	45	0.0	6.6
0.5 - 1.0	215	6	0.5	0.0
Combined	275	51	0.4	5.7

Tables XXI and XXII show that for the RSR site the differences in mean blood-lead levels and proportion of children with lead toxicity remain evident even when other sources of potential exposure (as determined from the questionnaire) were taken into consideration. Tables XXIII and XXIV show this is not the case for the Dixie site.

Table XXI. Percent Lead Toxicity by Distance, Traffic, and Other Potential Sources of Exposure for RSR Site

Approximate Distance (miles)	Traffic Density	Sample Size Potential Exposure YES	NO	Percent with Lead Toxicity Other Potential Exposure YES	NO
0.0 - 0.5	High	87	14	4.6	7.1
	Low	136	32	4.4	9.4
0.5 - 1.0	High	10	4	0.0	0.0
	Low	179	37	0.0	0.0

Table XXII. Percent Blood Lead $\geq$30 μg/dl by Distance, Traffic, and Other Potential Exposure for RSR Sites

Approximate Distance (miles)	Traffic Density	Sample Size Potential Exposure YES	NO	Percent with Lead 30 μg/dl Other Potential Exposure YES	NO
0.0 - 0.5	High	87	14	14.9	14.3
	Low	136	32	6.6	12.5
0.5 - 1.0	High	10	4	0.0	0.0
	Low	179	37	1.7	0.0

Table XXIII. Percent Blood Lead $\geq$30 μg/dl by Distance and Other Potential Exposure for Dixie Site

Approximate Distance (miles)	Sample Size Potential Exposure YES	NO	Percent Blood $\geq$30 μg/dl Potential Exposure YES	NO
0.0 - 0.5	74	32	4	0
0.5 - 1.0	134	90	3	0

Table XXIV. Percent Lead Toxicity by Distance and
Other Potential Sources of Exposure for Dixie Site

Approximate Distance (miles)	Sample Size Potential Exposure		Percent with Lead Toxicity Potential Exposure	
	YES	NO	YES	NO
0.0 - 0.5	74	32	4.3	-
0.5 - 1.0	134	90	.6	-

A multivariate analysis (Table XXV) shows the increased blood-lead level caused by the RSR smelter contribution and the traffic contribution to be 5.5 and 1.0, respectively.

Table XXV. Results of Statistical Test for Smelter
and Traffic Contribution for Blacks Only at RSR

Factor	Parameters		
	Percent with Lead Toxicity	Percent with Blood Lead $\geq 30\mu g/dl$	Mean Blood-Lead Level $(\mu g/dl)$
Smelter			
Contribution	5.7%	11.8%	5.5 $\mu g/dl$
Significance Level	.001	<.001	<.001
Traffic			
Contribution	.025%	3.7%	1.0 $\mu g/dl$
Significance Level	.494	.156	.130

Summary of Findings

1. At neither site is there evidence of absorption of lead to the degree usually associated with clinical symptoms of lead poisoning, and the reported blood-lead levels are not high enough to make this likely. However, a public health concern exists, particularly in the RSR area, since 5 percent of these black children were found to have lead toxicity.
2. At the RSR site, proximity to the smelter and to high traffic density contribute to the lead in the soil and to the blood-lead

level of children. In similar high traffic density areas, mean
blood-lead levels were 7.2 µg/dl higher near the site than in the
reference area. The contributing factors to the increased blood
lead appear to be residence in proximity to the smelter and, to
a lesser degree, proximity to high traffic density. There is a
resultant 5.6 percent of black children near the site with lead
toxicity as previously defined. This represents 14 of the 248
black children tested. This extrapolates to 29 of the 510
black children in the population. In comparing the RSR area
(within 0.5 mile) to the reference area (while holding traffic
constant), the smelter contribution to mean blood-lead levels
averages 6.6 µg/dl. In comparing the RSR areas of high and low
traffic (while holding smelter effect constant), the traffic
contribution to mean blood-lead levels averages 1.0 µg/dl. This
compares favorably with the contribution attributed to traffic in
the reference area of 1.4 µg/dl. Multivariate analysis shows the
ratio of smelter to traffic contribution to be 5.5 to 1.0 for the
RSR site.

3.   When the Dixie site is compared to the reference area, proximity
     to the smelter (within 0.5 mile) contributed 1.85 µg/dl to the
     mean blood-lead level. The potential contribution of traffic
     density could not be determined because of the configuration of
     the roadway and the distance of the roadway from the smelter
     site. Although an elevated mean blood-lead level was found for
     children living close to the Dixie site, the increase was not
     as great as observed in the RSR site; and the few children found
     to have lead toxicity, as defined previously, appear to have
     lead exposure due to occupation of parents.

RECEIVED September 12, 1984

An Approach to the Interdisciplinary Design of Multifactor Experiments

RALPH E. THOMAS

Battelle Memorial Institute, Columbus, OH 43201

The methodology described in this paper is intended to guide an inter-disciplinary team of scientists in developing an experimental design for a test program that can be characterized as a costly, time con-suming, complex multifactor experiment, possibly subject to high risks of failure. Lifetime tests and accelerated aging tests are examples. Complex environmental sampling programs could also be considered examples. The high risks of failure may be associated with insuffi-cient prior knowledge, an inability to control the experimental variables, synergisms, severe time and budget constraints, or a variety of other problems that may be uncovered once the design effort is underway. A formal well-documented team effort will help determine whether a test program should be implemented.

The proposed methodology is a specific procedure for designing a test program by extracting and synthesizing expert opinion. Because the envisioned tests are costly, it is important to assign responsi-bility for the experimental design and documentation that ultimately support or refute the implementation of the test program to a compe-tent team of individuals. The procedure differs from similar methods in its reliance on certain basic concepts related to the statistical design of experiments.

The team members must represent the various scientific disci-plines that are associated with the experimental processes of in-terest. The methodology requires the design team to meet several times over a period of several weeks to carry out a set of formal agendas described below and to conclude its efforts with a formal document. The document will contain either a detailed experimental design that meets the constraints of the envisioned test program, or the document will show that an acceptable experimental design does not exist and the envisioned experiment should not be implemented. In either case, the detailed documentation permits subsequent peer review.

An experimental design is sometimes taken to be a "test matrix" that indicates the conditions associated with each test measurement. However, for statistical purposes the test matrix must reflect both the "experimental design" and the "data analysis" characteristics that are required to identify separately the magnitudes of the different

0097-6156/84/0267-0067$06.00/0

effects and their synergisms. In short, the purpose, the model, and the method for analyzing the data must be specified in advance, and together they constitute the experimental design.

Because of the emphasis on experimental design, it is required that a statistician serve as a member of the design team. The assigned tasks and responsibilities for the statistician differ from those for the scientists. The primary mechanism for obtaining the experimental design is to require each scientist on the team to make explicit, documented, numerical predictions for all combinations of the test conditions specified in a factorial table. In effect, such predictions require each scientist to quantify the effects of the experimental factors (control variables) on the dependent variable. These predictions are based on the scientist's knowledge and assessment of related literature, data, experience, etc. Candidate team members who are unable or unwilling to make such predictions are excluded from the team.

The predicted results are analyzed in terms of main effects and interactions by the team statistician using standard statistical methods. The results are presented in a graphical form using a hierarchical tree which aids conceptualization. They are also described using at least one mathematical model. Each scientist is required to iterate among the factorial table, the hierarchical tree, and the mathematical model until all three forms are equivalent and correctly represent the scientist's expectations. When this stage is reached in the methodology, the team scientists then compare their expectations. This is done primarily by mutual examinations of the hierarchical trees and the associated mathematical equations. Differences in the trees and equations are next discussed by the team members with reference to available documentation (literature, data, experience, etc.). At this stage the objective of the team is to identify the most defensible documentation and arrive at a consensus factorial table, hierarchical tree, and mathematical model.

The final stages of the methodology consist of eliminating some of the less informative tests and adding other tests to provide data needed to quantify nonlinear relationships. The resulting experimental design then consists of a mix of conditional main effects and conditional interactions. It represents a compromise among scientific, statistical, and economic constraints.

The team activities are coordinated and controlled by a team manager. The team manager must have sufficient authority to arrange meetings, make assignments, focus discussions, limit debates, and ensure that good documentation is obtained. The required effort is appreciable and may range between four and twelve man-months for a team of five to ten members. The approach is recommended when firm, supporting documentation is necessary.

The methodology described in this paper has evolved over the past ten years but still remains very elementary. It was developed to minimize the exceptional hazards associated with the design and implementation of accelerated life tests. Many accelerated life tests can be characterized as expensive failures, due to the use of poor experimental designs, inadequate scientific and statistical expertise, and insufficient peer review prior to implementation. The methodology outlined below is intended to reduce such failures.

In the area of accelerated life testing, applications of the
recommended methodology have been made to spacecraft batteries (1)
and to photovoltaic cells (2) (3). Applications are currently being
made to accelerated life testing of insulation materials used in high
voltage cables, weapons structure, and the packaging of radioactive
wastes. Brief descriptions of the methodology in the accelerated
test setting are given in References (4), (5), and (6).

The Team Approach

The time schedules for accomplishing the following agendas are usually
decided by the team manager. The items on each agenda provide general
guidance. Each team modifies the agendas to accommodate the unique
requirements of the system under consideration and to provide for
preparing and assessing the factorial tables, hierarchical trees,
associated mathematical models, and related documentation.

A preliminary meeting is usually required to describe the
approach to candidate members of a design team. At this stage the
responsibilities and obligations of the team members are made sufficiently clear so that prospective team members can evaluate their
willingness and ability to participate.

Agenda #1. The first formal meeting of the team is based on the
following agenda:

- o Identify the experimental factors that define the test
  conditions.
- o Define the dependent variables.

In general, this agenda requires at least one day and allows
each team member to present relevant background and views concerning
the proposed experiment. The identification of the test conditions
and experimental means for controlling them usually requires knowledge
of experimental procedures associated with controlled environments,
monitoring, and instrumentation. Initial estimates of experimental
costs may also be introduced at this first meeting.

The team must have full responsibility for choosing both the test
factors and the dependent variables. It is usually necessary to repeat the procedure described below for each dependent variable. Examples of dependent variables selected by teams include the following:
relative "severity" of each test, percentage loss in a performance
measure relative to the initial level of performance, and percentage
change in a performance measure resulting from a percentage change in
a test condition. Examples of test factors include temperature, pH,
relative humidity, and radiation.

It is essential that good documentation procedures be established
during the first meeting of the team.

Agenda #2. The second meeting of the team is based on the following
agenda:

- o Obtain a group consensus on suitable test ranges for each
  factor by specifying a low and high level for each factor.

For notational convenience, the low and high levels of each factor are denoted by L and H, respectively.

o Form a complete factorial table consisting of all combinations of low (L) and high (H) values for each factor. If there are n factors being considered, then the complete factorial table will consist of the 2^n possible combinations. This list of combinations of factor levels provides a guide for obtaining the information necessary to generate the experimental design.

o Make a preliminary assessment of each of the combinations of low and high levels for the purpose of verifying that the individual combinations of levels are suitable for experimental implementation. Some combinations may not be technically feasible. If some combinations are not acceptable, revise the ranges of the factors until each combination of levels in the 2^n factorial table can be accepted as a suitable test condition.

o For a selected dependent variable and for each of the 2^n possible test conditions, have each scientist provide an estimate (prediction) of the numerical value of the dependent variable that would be expected if the test combination were included in the final experimental design. This assignment is usually the most difficult task the individual scientist is required to perform. Because of the difficulty, the task is typically continued as a week-long assignment to permit each scientist to assemble data, refer to literature, examine previous experimental results, etc.

Agenda #3. This agenda includes:

o An analysis is conducted of the predicted values for each team member's factorial table to determine the main effects and interactions that would result if the predicted values were real data. The interpretations of main effects and interactions in this setting are explained in simple computational terms by the statistician. In addition, each team member's results are represented in the form of a hierarchical tree so that further relationships among the test variables and the dependent variable can be graphically illustrated.

o The team statistician then discusses the statistical analysis and the hierarchical tree representation with each team scientist.

o Each participant is permitted to revise the predictions until satisfied that both the factorial table (which focuses on combinations of the test variables) and the associated hierarchical tree (which focuses on the individual test variables) properly reflect the scientist's views concerning the anticipated relationships among the test conditions and the predicted values of the dependent measure.

o The trees for each participant are next compared and synthesized to obtain a final group consensus. This means that a single factorial table and associated tree must be identified. The synthesis process tends to expose and highlight conflicting views regarding the anticipated relationships among the test variables. These views are recorded and serve as part of the documentation of the design process.

The objective of this agenda is to obtain a consensus factorial table, an associated hierarchical tree, and a mathematical model that generates the predicted values of the dependent variable and that is supported by documented consensus arguments, data, and calculations. The process of obtaining a consensus can be difficult. However, most scientists seem to enjoy comparing arguments, data, and models with a view toward identifying the best overall compromise. The manager of the team is responsible for ensuring that an acceptable compromise is obtained.

Agenda #4. The next step in developing the experimental design involves "pruning" the tree. It is usually found that the consensus hierarchical tree calls for an excessive number of tests. For example, some tests may have been run already or it may be that, under certain conditions in certain branches of the tree, the effect of changing a factor from its low to high level would be expected to cause insignificant changes in the value of the dependent variable. Consequently, all splits in the hierarchical tree that are associated with relatively small changes are examined. These correspond to conditional main effects that are expected to be small and can possibly be eliminated from the final test design.

The elimination of tests that results from pruning must be accompanied by careful documentation that explains the basis for the elimination. It is frequently desirable to account for the budget constraints at this time. If the tree cannot be pruned to a level consistent with the budget, the arguments of the team should be documented to indicate that either no test should be implemented or that a specified budget increase is necessary.

A complete factorial design can be severely degraded by pruning. It may no longer be possible to obtain satisfactory estimates of certain main effects and interactions. For this reason the team statistician is next required to assess the statistical properties of the test design associated with the pruned hierarchical tree. If the design is statistically unacceptable, the statistician is charged with revising the design to achieve acceptability. In simple cases the revision may be accomplished by adding back a minimal number of tests previously eliminated by the team scientists. In more complex cases the statistician may choose to deviate from the factorial basis of the design and focus on other designs (exploratory, sequential, etc.) that are recognized as appropriate and consistent with the information generated by the team. The revised experimental design must be documented by the statistician for review by the team scientists as a part of Agenda #5.

Agenda #5. The fifth agenda calls for the team scientists to accomplish the following:

o Evaluate the statistician's recommended design.
o Arrive at a possibly revised consensus design.
o Review the overall design and insert additional control levels between the low and high levels, if necessary, to provide for anticipated nonlinear relationship.

○ Re-examine the final test conditions to ensure that, when
  implemented, the test data will be sufficient to estimate all
  parameters in the fitted mathematical models.
○ With the statistician's participation, identify the number of
  replicates to be made at each test condition.
○ Identify the test instrumentation and measurement procedures,
  schedules, etc., that are required for a fully specified test
  design.
○ Document all choices with the reasons that support each
  choice.

Agenda #6. The last agenda consists of a team review and approval of
a write-up that documents the final test design. The documentation
must include the consensus factorial table, hierarchical tree, and
mathematical model used to fit the predicted values. In addition,
the documentation must include all basic arguments and considerations,
even if these considerations do not appear in explicit form in the
final design. The specific reasons for excluding certain test
conditions, certain test variables, etc., must be included in the
documentation. For subsequent reviews of the proposed experimental
design, the reasons that underlie what is not recommended are sometimes as important is those that underlie what is recommended.
   Ideally, the design team is responsible for making concurrent
analyses of the actual test data as the data become available. The
team's understanding of the anticipated relations can be established
by requiring the team scientists to make real time predictions of two
kinds during the course of the experiment. One kind occurs "within"
a test condition. In this case the data obtained to date from a
particular test condition are used to predict the measurement values
to be expected at a specified future date, say a week or month later
for the same test condition. A second kind of prediction involves
"across" test conditions. In this case the data obtained at one test
condition are used to predict the measurement values for a different
test condition at future specified dates.

An Example in Applying the Approach

Factorial tables, hierarchical trees, and associated mathematical
models are elementary tools used to guide the efforts of the design
team.

Factorial Tables. Table I shows a factorial table in a general form
for three experimental factors. The factors are denoted by X_1, X_2,
and X_3, and each factor is considered to be imposed either at a low
level, symbolized by L, or at a high level, symbolized by H. For
three factors, there are eight (2^3) combinations of the low and high
stress levels. These combinations are shown in a standard order as
(L,L,L) through (H,H,H) in Column 2. Each of the eight combinations
represents a possible test condition. The table can be easily extended or reduced to accommodate n factors with 2^n combinations of high
and low levels. The n factors to be used in the test must be defined
by the design team together with the low and high levels for each
factor.

Table I. Factorial Table To Be Independently Completed
by Each Scientist on the Test Design Team
(General Form for 3 Factors)

Test Number	Test Combination of Levels (1) (X_1, X_2, X_3)	Predicted Value of Y	Documentation Supporting the Predicted Value
1	(L,L,L)	Y_1	(1)
2	(L,L,H)	Y_2	(2)
3	(L,H,L)	Y_3	(3)
4	(L,H,H)	Y_4	(4)
5	(H,L,L)	Y_5	(5)
6	(H,L,H)	Y_6	(6)
7	(H,H,L)	Y_7	(7)
8	(H,H,H)	Y_8	(8)

(1) L, H denote low and high levels for factors X_1, X_2, and X_3.

Table II shows, as an example, the combinations of low and high levels for three factors selected by a design team for an accelerated test involving photovoltaic solar cells. In column 2 the three factors are seen to be temperature T (50°C, 95°C), relative humidity RH (60%, 85%), and ultraviolet radiation UV (five suns, 15 suns). The eight combinations of the high and low levels are shown, together with the predicted months to failure for each combination. In this example the documentation to support each prediction is symbolically referenced as shown in the last column. The documentation includes assumptions, calculations, references to the literature, laboratory data, computer simulation results, and other related material. Such a factorial table is first completed by each scientist independently. Subsequently, the team aims to generate a single consensus factorial table has the same form as that shown in Table II.

Hierarchical Trees. The hierarchical tree is constructed by the team statistician for each scientist in accord with the factorial table filled out by the scientist. To be useful for this purpose the tree must have the capability of exhibiting virtually any conceivable relation among the test factors and the dependent variable. It would be undesirable if the scientists were forced to constrain the anticipated relationships in any manner.

Figure 1 shows the hierarchical tree that corresponds to Table I. The scale at the top of the tree corresponds to the predicted time to failure in months. The conditions at the top of the tree are given by 71°C, 72%, and 8.3 suns and correspond to an expected failure time of approximately 16 months. The first split in the tree is associated with a temperature of 95°C on the left branch and 50°C on the right branch, which correspond to average predicted lifetimes of approximately four and 27 months, respectively. The next two splits are seen to be based on relative humidity and ultraviolet radiation.

Table II. Example of a Completed Factorial Table

Test Number	Test Combination of Levels (T°C, RH%, UV suns)	Predicted Time to Failure, months	Documentation Supporting Predicted Value
1	(50, 60, 5)	40.0	(1)
2	(50, 60, 15)	31.0	(2)
3	(50, 85, 5)	26.5	(3)
4	(50, 85, 15)	11.9	(4)
5	(95, 60, 5)	8.0	(5)
6	(95, 60, 15)	6.2	(6)
7	(95, 85, 5)	4.6	(7)
8	(95, 85, 15)	3.6	(8)

The first split of the hierarchical tree partitions the eight rows of data in Table II into two sets of four rows each, with four rows corresponding to a temperature of 50°C and the remaining four rows corresponding to a temperature of 95°C. The splitting process for the tree continues in a sequential manner, with each split further partitioning the data into two subsets that correspond to the low and high values of the splitting variable. The splitting process terminates when the ends of the branches correspond to the individual test conditions associated with each row of Table II.

Temperature is used at the first splitting variable in Figure 1 because numerical calculations show that temperature is a better predictor of life than either relative humidity or ultraviolet radiation at this stage. For both the low and high temperature branches of the tree, the numerical calculations show that the second most important predictor is relative humidity. Because no other variables remain, the final splits are necessarily based on ultraviolet radiation.

Figure 2 illustrates the variety of forms that can occur for hierarchical trees. Figure 2(a) shows a symmetrical tree for two factors X_1 and X_2. No interaction between X_1 and X_2 is said to occur if the horizontal distances associated with X_2 are approximately equal at the low and high levels of X_1 as shown in Figure 2(a). The presence of a strong interaction between X_1 and X_2 is shown in 2(b), where the horizontal distances associated with X_2 at the low and high levels of X_1 are markedly unequal. In this case, the variable X_2 has a relatively large effect at the high level of X_1 and a relatively small effect at the low level of X_1. Figure 2(c) shows a tree in which the second stage predictor is X_2 when X_1 is at its low level. However, a different predictor X_3 is indicated when X_1 is at its high level. Finally, Figure 2(d) shows a tree in which an increase in X_2, from low to high, has opposite effects, depending on whether X_1 is at its high or low level.

To date every relationship anticipated by team scientists has been easily represented by a hierarchical tree. Moreover, all of the structures shown in Figure 2 have actually occurred in applications. Thus, it is not uncommon for interactive synergisms to be

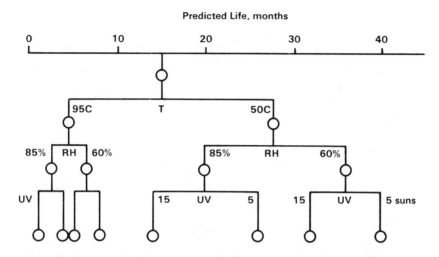

Figure 1. Hierarchical tree showing predicited life as a function of test variables.

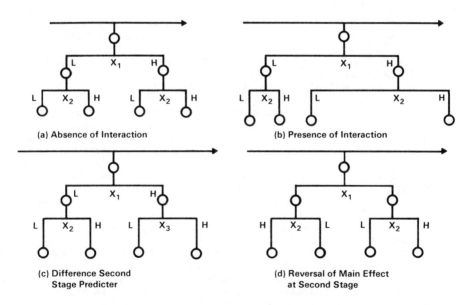

Figure 2. Different types of hierarchical trees.

anticipated, especially when temperature is one of the experimental factors. Further, the more important variables involved in one branch of a tree may well differ from those in another branch. Finally, but more rarely, reversals in the direction of main effects also occur. For these reasons the hierarchical tree and the associated factorial table are believed to have the capability of properly representing relationships anticipated by the team scientists.

Comparisons among the trees immediately identify the areas of agreement and disagreement among the scientists. Differences in the sequence of dependent variables down each branch, differences in horizontal displacements, and reversals in the directions of various main effects are important indicators of major disagreements among the scientists. Such information is difficult to obtain by direct visual inspections of the factorial tables or by verbal communication. For the first iteration the team statistician is responsible for extracting the information by computational methods, and then presenting the results for each table in the form of a hierarchical tree.

Pruning the Tree. In many practical cases the complete factorial design requires too many tests. In such cases the design team is permitted to eliminate, at least provisionally, those test combinations believed to be less essential. Such test combinations are usually associated with small horizontal separations in the lower branches of the hierarchical tree. When these branches are deleted, the hierarchical tree is said to be "pruned". The team statistician is responsible for assessing the "damage" associated with pruning the tree and with revising the experimental design to achieve acceptability. It is anticipated that the team statistician may reintroduce certain tests or may completely revise the experimental design to best conform with the infomation that has been generated during the course of design process. In either case, the revision of the experimental design requires a reappraisal of the test matrix and the associated mathematical models that describe the predicted results.

Mathematical Models. As noted previously, a mathematical model must be fitted to the predicted results shown in each factorial table generated by each scientist. Ideally, each scientist selects and fits an appropriate model based upon theoretical constraints and physical principles. In some cases, however, appropriate models are unknown to the scientists. This is likely to occur for experiments involving multifactor, multidisciplinary systems. When this occurs, various standard models have been used to describe the predicted results shown in the factorial tables. For example, for effects associated with lognormal distributions a multiplicative model has been found useful. As a default model, the team statistician can fit a polynomial model using standard least square techniques. Although of limited use for interpolation or extrapolation, a polynomial model can serve to identify certain problems involving the relationships among the factors as implied by the values shown in the factorial tables.

The form of the mathematical model fitted to the consensus factorial table must be reassessed after the hierarchical tree is pruned and the experimental design has been revised by the statistician.

The reassessment is necessary to ensure that sufficient data are to be obtained to permit the estimation of all parameters.

Literature Cited

1. "Accelerated Testing of Space Batteries," National Aeronautics and Space Administration SP-323, 1973.
2. "Methodology for Designing Accelerated Aging Tests for Predicting Life of Photovoltaic Arrays," Final Report, Energy Research and Development Agency/Jet Propulsion Laboratory-954328-77/1, 1977.
3. "Development of an Accelerated Test Design for Predicting the Service Life of the Solar Array at Mead, Nebraska," Final Report, Department of Energy/Jet Propulsion Laboratory-954328-79/13, 1979.
4. Thomas, R. E.; Gaines, B. G. "Methodology for Designing Accelerated Aging Tests for Predicting Life of Photovoltaic Arrays"; Department of Energy/National Bureau of Standards Workshop on Stability of Thin Film Solar Cells and Materials, Washington, D.C., May 1-3, 1978.
5. Thomas, R. E.; Gaines, G. B. "Procedure for Developing Experimental Designs for Service Life Prediction"; Thirteenth Institute of Electrical and Electronics Engineers, Inc. Photovoltaic Specialists Conference, Washington, D.C., June 5-8, 1978.
6. Thomas, R. E.; Gaines, G. B.; Epstein, M. M. "Methodology for Estimating Remaining Life of Components Using Multifactor Accelerated Life Tests"; Proceedings of the Workshop on Nuclear Power Plant Aging, Bethesda, Maryland, August 4-5, 1982.

RECEIVED August 6, 1984

9

Statistical Methods in Environmental Sampling

Radian Corporation, Austin, TX 78766

This paper discusses the role that statistics can play in environmental sampling. The primary difference between an investigation based on statistical considerations and one that is not is the degree of objectivity that can be incorporated into the evaluation of the quality and uncertainty of the study results. Statistical methods in the planning stage can also aid in optimizing allocation of resources.

Most environmental sampling studies are not amenable to classical statistical techniques. Correlation among samples, non-normal distributions of measurements, and multivariate requirements are typical in environmental studies. The effective use of statistics in an environmental study thus depends on meaningful interaction between statisticians and other environmental scientists.

Sampling studies can be classified into two types – enumerative, or descriptive, and analytic (1). The classification is important because the applicable statistical methods and approaches are different for these two types. The objective of either type of study is to provide a basis for action. In an enumerative study the action is directed to the population from which the samples were taken. How or why the population was formed is not of primary interest. In an analytic study, the primary interest is the causal system or process which created the conditions observed in the study. Action taken is directed toward this process rather than the population sampled.

Environmental sampling studies include both the enumerative and the analytic types. Some studies provide data for both enumerative and analytic action. Examples of enumerative studies include testing of soil or wastes to define required cleanup, measuring emissions from process sources to identify those requiring maintenance, and determining locations of air monitoring stations to monitor pollutant levels. Examples of environmental analytic studies include evaluating new control technologies, sampling streams and lakes to identify sources of acid rain, and validating analytical methods. Typically, an environmental assessment is an enumerative study while environmental research studies are analytical in nature. This paper addresses statistical methods for enumerative environmental sampling studies.

0097–6156/84/0267–0079$06.00/0
© 1984 American Chemical Society

Steps in Environmental Studies

Environmental sampling studies are extremely variable in scope, duration, and complexity. The following steps are general to this wide variety of studies.

Objectives of the Study. A clear statement of the objectives of the study is required if statistical methods are to be used effectively in study planning. Many studies have multiple objectives which compete for study resources. An understanding of these objectives by all involved parties at the outset usually leads to better studies.

From a statistical viewpoint, there are some important specifics in study objectives for an enumerative study (2). Objectives can be categorized into the following three groups:

1. to estimate or evaluate the average of characteristics of interest in the population;
2. to estimate or evaluate the variability or distribution of characteristics of interest in the population;
3. to decide if characteristics of interest in the population meet certain standards or pre-established criteria.

Statistical methods used for planning will depend on which of the above three objectives is important.

Study Population. A description of the materials, areas, industry, etc., to be studied is required. The population denotes the aggregate from which samples are selected (3). Ideally, the sampled population should coincide with the population for which information is required (the target population). Sometimes, due to practical constraints or for convenience, the sampled population includes only a portion of the target population. For example, in studying hydrocarbon fugitive emissions, the target population in a refinery might be all in-process valves while the sampled population is all valves which can be tested from ground level. Statistical inference from such a study will be to the sampled population. Judgment of substantive experts is required to extend the inferences to the target population.

Characteristics of Interest and Methods of Measurement. Each characteristic of the population to be measured or observed should be stated along with a method measurement. Characteristics can be grouped into three types of data:

1. discrete measurements - measurements are either classifications or categorizations (e.g. presence or absence of a pollutant, weather conditions such as raining or not raining, etc.);
2. continuous measurements - measurements are made on a continuous scale (e.g. pollutant concentration trends, temperature or rainfall in 24-hour periods, etc.);
3. mixed measurements - mixture of discrete and continuous measurements (e.g. concentration of pollutant if greater than detection limit; otherwise reported as not detected).

Prior information concerning expected magnitudes, variability, and distributions of measurements of each characteristic can be used in developing an efficient sampling program. Information on the measurements such as between-laboratory and within-laboratory bias and precision are important in planning the sampling strategy.

<u>Degree of Precision Required.</u> A statement of the required precision of the results from the sampling effort for each characteristic of interest is needed to allocate sampling and testing resources efficiently. As previously discussed, the form of these statements will depend on the objective and type of characteristics. The users of the study results must make these determinations. Statistical methods to evaluate tradeoffs and alternatives may be useful in assisting the study administrators in this effort.

The use of confidence intervals is one way to state the required precision. Confidence limits provide a measure of the variability associated with an estimate, such as the average of a characteristic. Table I is an example of using confidence intervals in planning a sampling study. This table shows the interrelationships of variability (coefficient of variation), the distribution of the characteristic (normal or lognormal models), and the sample frequency (sample sizes from 4 to 365) for a monitoring program.

<u>Design of the Sampling Study.</u> When the objectives, populations of interest, characteristics to be determined, and required precision are known, the sampling study can be designed. The design should include the following elements:

1. partitioning of the sampling population into subpopulations or strata;
2. division of the sampling population (or strata) into sampling units;
3. determination of the number and type of sampling units to be collected or tested;
4. procedures for selecting and obtaining particular sampling units (4) including sampling equipment, selection procedures, and temporal considerations;
5. sample handling procedures (compositing; sub-sampling; pretreatment such as filtering, drying, and sieving; chain-of-custody; and quality assurance procedures);
6. data collection forms for recording observations and pertinent sampling information;
7. procedures for summarizing and analyzing the results of the study

The design should be summarized in a format to allow review and revisions prior to implementation. Required resources and schedules should be included in this document.

<u>Conduct of the Sampling Study.</u> It is often desirable to conduct an environmental sampling study in stages, with laboratory analysis and data analysis of the first stage completed prior to subsequent sampling pling stages. Modifications to objectives, characteristics of interest, and study design then can be made after the first stage. This sequential experimentation process is discussed later.

Table I. Expected Confidence Intervals for a Parameter Mean as a Function of Number of Samples (Measurements)

Expected Variability of Measurement (Coefficient of Variation)[a]	Distribution (Model)	95% Confidence Interval About the Mean Estimate (percent)[b]					
		n = 4 (Quarterly)[c]	n = 6 (Bi-monthly)[c]	n = 12 (Monthly)[c]	n = 24 (Semi-monthly)[c]	n = 52 (Weekly)[c]	n = 365 (Daily)[c]
5%	Normal Model	+8.0	+5.2	+3.2	+2.1	+1.4	+0.5
	Lognormal Model	+8.1,-7.5	+5.4,-5.1	+3.2,-3.0	+2.1,-2.0	±1.4	±0.5
10%	Normal Model	+15	+11	+6.4	+4.2	+2.8	+1.0
	Lognormal Model	+17,-15	+11,-9.9	+6.5,-6.1	+4.3,-4.1	+2.8,-2.7	±1.0
25%	Normal Model	+40	+26	+16	+11	+6.9	+2.6
	Lognormal Model	+48,-32	+29,-23	+17,-14	+11,-9.9	+7.1,6.6	+2.6,-2.5
50%	Normal Model	+80	+52	+32	+21	+14	+5.2
	Lognormal Model	+110,-53	+62,-39	+35,-26	+22,-18	+14,-12	+5.0,-4.8
100%	Normal Model	+160	+110	+64	+42	+28	+11
	Lognormal Model	+280,-73	+104,-58	+70,-41	+42,-30	+26,-21	+9.0,-8.2
200%	Lognormal Model	+650,-87	+280,-74	+124,-55	+71,-42	+43,-30	+14,-12
500%	Lognormal Model	+1700,-94	+570,-85	+220,-68	+120,-53	+65,-40	+20,-17
1000%	Lognormal Model	+3000,-96	+900,-90	+300,-75	+150,-60	+82,-45	+24,-20
10000%	Lognormal Model	+13000,-99	+2300,-96	+590,-85	+260,-72	+130,-57	+37,-27

[a]Coefficient of variation is the ratio of the standard deviation to the mean expressed in percent. A known value for the coefficient is assumed in this table. If the coefficient of variation is greaterthan 100%, the normal model is not realistic.

[b]n = number of samples or data points.

[c]Monitoring frequencies required for the specific value of n if the duration of study was one year.

Many characteristics of interest in an environmental sampling study require chemical or physical analysis in a laboratory. When a laboratory is involved, the sampling design must include consideration of sample processing in the laboratory and analytical protocols.

When testing and analysis are completed, the data can be analyzed and summarized. Statistical methods are often used during this step in a study. Data should first be edited and validated. Quality assurance information from both the sampling and laboratory analyses should be considered in this validation. Field sampling personnel and laboratory scientists should maintain responsibility for data validation.

The data summarization procedures will depend on the objectives and type of data. Statistical calculations should be supported with graphical analysis techniques. A statement of precision and bias should be included with all important results of the study.

Sampling Models

A statistical sampling model is a mathematical representation of a sampling study. Models are especially useful in studying variability of study estimates and sources of variability. Models include both physical aspects of the sampling study and theoretical statistical considerations. Certain assumptions are required when using the statistical models. One assumption is that samples are selected in an unbiased and independent manner from the sample population (or each strata in the population). This assumption can usually be assessed by employment of random sampling procedures in the selection of sample units. The use of sampling models to maximize the amount of information obtained in a study for a given cost is probably the most important contribution of statistical methods to sampling problems.

The simplest model arises when sampling units are randomly selected from a large target population and analyzed without analytical error. If the objective of the study is to estimate the average concentration of a pollutant in a population (letting x represent the concentration, a continuous variable), then

$$\sigma_{\bar{x}}^2 = \sigma_s^2 / s \qquad (1)$$

where $\sigma_{\bar{x}}^2$ = variance of the average concentration estimate

σ_s^2 = variance of the sampling units, and

s = number of samples selected.

If analytical variation is important, then

$$\sigma_{\bar{x}}^2 = \sigma_s^2 / s + \sigma_a^2 / as \qquad (2)$$

where σ_a^2 = variance of the analytical protocol, and

a = number of replicate analyses per sample.

The terms σ_s^2 and σ_s^2 are called variance components. More complicated sampling schemes involve stratification of the population, compositing, subsampling for analysis, and between-laboratory and within-laboratory sources of variability; and these factors add additional components of variation to the model for the variance of the average concentration.

Another complication is a finite population correction which takes the form of

$$[(N-n)/N]\sigma_i^2 \qquad (3)$$

where n represents the number of samples selected and N is the total number of samples in the population. An example in which this correction factor might be important would be sampling wastes stored in barrels. N is the total number of barrels in the population (i.e. the waste site), and n is the number of barrels sampled. σ_i^2 is the between-barrel variance component. In most environmental sampling problems the population is large enough to ignore the finite population correction factor.

Figure 1 gives some models for a variety of environmental sampling situations when estimating the average of a characteristic is the relevant objective.

The usefulness of these models in planning an environmental sampling program depends on the availability of estimates of the variance components for the important sources of variability. The following two sections describe the use of these models in selecting sample sizes, designing sampling studies, and allocating resources. Estimates for some of the components, such as analytical variability, are often available. Other components must be estimated from previous studies or pilot studies. A general two-stage procedure for estimating components is described later.

Determining Sample Size

The most common question posed to statisticians in environmental sampling is "How many samples do I need to take?" (or "How many replicates," "How many analyses," etc.). The statistical models introduced previously provide a framework for addressing these questions after the first four steps in a sampling study are completed (i.e. the objectives, populations of interest, characteristics to be determined, and required precision are stated). The methods in this section are applicable when the objective is to estimate the average of a characteristic in the population.

One measure of the quality of an estimate of an average is the confidence limits (or maximum probable error) for the estimate. For averages of independent samples, the maximum probable error is

$$E = Z_p \, (\sigma_{\bar{x}}) = \frac{Z_p \sigma}{\sqrt{n}} \qquad (4)$$

where n is the number of samples, σ is the standard deviation of the sample measurements, and Z_p is a percentile of the standard normal distribution. The equation is exact if test results are normally

distributed and σ is known.   If results are not normally distributed,
the equation provides an approximation which improves as n increases.
P represents the probability associated with the confidence limit.
For P = 95%, z_p = 1.96.   Equation 4 can be rearranged to give the
number of tests required for an estimator with specified maximum
probable error:

$$n = z_p^2 \sigma^2 / E^2 \qquad (5)$$

For example, if a maximum error of 5 ppm is desired for the average
concentration of a particular chemical with 95% confidence and the
standard deviation of sample results is 10 ppm, then

$$n = \frac{(1.96)^2 \ (10)^2}{5^2} = 15.4$$

and 16 samples would be required.
   If the variability (σ) depends on concentration, prior knowledge of concentration may be required to use these formulas.   If the
relative standard deviation (RSD) is constant with respect to concentration, then the formulas can be applied by interpreting σ and E as
relative standard deviation and relative error, respectively.   A common case in which RSD is constant with respect to concentration is
when analytical results are lognormally distributed.   For example,
suppose it is desirable to estimate the average concentration with
95% confidence that the estimate will be within 10% of the true value
if the relative standard deviation is 25%.   Then

$$n = \frac{1.96^2 \ (25)^2}{10^2} = 24 \qquad (6)$$

and 24 samples would be required.
   Equation 5 can be evaluated using the nomograph in Figure 2.
For example, to find the n needed to achieve a maximum error of 5 ppm,
as in the first example above, first find the point where the diagonal
intersects the line through E = 5 ppm and P = 95%.   Then the line
through this point and σ = 10 ppm cuts the n scale at the required
value, n = 16.
   Equation 4 also can be evaluated using the nomograph.   For
example, to determine the maximum probable error that will occur with
95% probability based on n = 4 tests when σ = 20 ppm, first find the
point where the diagonal and the line through n = 4 and σ = 20 inter-
sect; then extend the line through this point and P = 95% to find E =
19.6 ppm.
   The n value obtained using equation 5 will sometimes be infea-
sible for economic reasons.   In such cases, the nomograph facilitates
finding E and P combinations that yield a practical n.   Particular
choices of n can be evaluated by finding the diagonal point on the
line connecting n and σ, then finding E and P values on lines through
this point.   It can be seen from the nomograph that n increases with
increasing confidence level (P) or decreasing error (E).   For fixed

n and σ, smaller error requirements mean that a lower confidence level must be accepted.

All factors in the equation except σ can be varied by the sampling study designer. The standard deviation is characteristic of the variability of the samples in the population.

The simplest of sampling models (a single component of variation) was used in constructing the nomograph in Figure 2. Equations similar to 4 and 5 can be developed for the more complex models discussed previously. For example, a model with two components (i.e. sampling and analytical variation, or two-stage sampling) would result in average estimates with maximum probable error

$$E = Z_p \left[\frac{\sigma_1^2}{n_1} + \frac{\sigma_2^2}{n_1 n_2} \right]^{1/2} = Z_p(\sigma_{\bar{x}}) \qquad (7)$$

where σ_1^2 is the variance component for the first stage, σ_2^2 is the variance component for the second stage, and n_1 and n_2 the number of the first and second stage samples (or analyses), respectively. Either n_1 or n_2 or both can be increased to obtain the desired value for E. Equation 7 takes on a similar form for the other sampling models discussed in the previous section. Consider the two-stage sampling model with analytical variation. Then

$$\sigma_{\bar{x}} = \frac{\sigma_c^2}{c} + \frac{\sigma_s^2}{cs} + \frac{\sigma_a^2}{csa}$$

and 4.1 becomes

$$E = Z_p \left[\frac{\sigma_c^2}{c} + \frac{\sigma_s^2}{cs} + \frac{\sigma_a^2}{csa} \right]^{1/2} \qquad (8)$$

where σ_c^2, σ_s^2, and σ_a^2 are variance components for containers, samples, and analytical variability and c is the number of containers sampled, s in the number of samples per container, and a is the number of analyses per sample.

A maximum probable error of E can be obtained in a variety of ways. If $\sigma_c^2 = 10\%$, $\sigma_s^2 = 40\%$, and $\sigma_a^2 = 20\%$, then Table II shows some possible sampling schemes to obtain E less than 3%. The choice among alternate schemes is discussed in the next section.

Several complications can occur in environmental sampling that require extensions of the methods discussed above. These include:
 ○ correlations between samples or analyses;
 ○ measurement of several characteristics on the same sample;
 ○ use of composite sampling procedures in place of arithmetic averages.

Correlations, such as commonly exist between hourly air pollution measurements or wastewater samples taken on successive days, violate the key assumption of the equations in this section. When such correlations exist, alternate methods must be used (5,6). When several characteristics are measured on a sample, the methods of this section can be applied separately for each characteristic. If results for different characteristics conflict, one can pick the result that works best for all parameters or the result for the most important

Model Description	Model Error	Definitions of Terms
A) No Analytical Variation:		$\sigma_{\bar{x}}^2$ = variance of average
1) Simple Random Sampling	$\sigma_{\bar{x}}^2 = \dfrac{\sigma_s^2}{s}$	σ_s^2 = sample variance component
2) Two-Stage Sampling (Strata and Samples)	$\sigma_{\bar{x}}^2 = \dfrac{\sigma_c^2}{c} + \dfrac{\sigma_s^2}{cs}$	σ_c^2 = strata variance component
3) Three-Stage Sampling (Strata, Samples, Subsamples)	$\sigma_{\bar{x}}^2 = \dfrac{\sigma_c^2}{c} + \dfrac{\sigma_s^2}{cs} + \dfrac{\sigma_b^2}{csb}$	σ_b^2 = subsampling variance component
4) Compositing Samples	$\sigma_{\bar{x}}^2 = \dfrac{\sigma_s^2}{s} + \dfrac{\sigma_r^2}{r}$	σ_r^2 = variance component for compositing procedure
B) Analytical Variation:		σ_a^2 = within-laboratory analytical variance component
1) Simple Random Sampling, Within-Lab Variation	$\sigma_{\bar{x}}^2 = \dfrac{\sigma_s^2}{s} + \dfrac{\sigma_a^2}{sa}$	σ_e^2 = between-laboratory analytical variance component
2) Simple Random Sampling, Between-Lab Variation	$\sigma_{\bar{x}}^2 = \dfrac{\sigma_s^2}{s} + \dfrac{\sigma_e^2}{e} + \dfrac{\sigma_a^2}{sa}$	s = number of samples
3) Compositing Samples, Within-Lab Variation	$\sigma_{\bar{x}}^2 = \dfrac{\sigma_s^2}{s} + \dfrac{\sigma_r^2}{r} + \dfrac{\sigma_a^2}{sa}$	c = number of strata sampled
4) Composite Samples; Subsampling Within-Lab Variation	$\sigma_{\bar{x}}^2 = \dfrac{\sigma_s^2}{s} + \dfrac{\sigma_r^2}{r} + \dfrac{\sigma_b^2}{rb} + \dfrac{\sigma_a^2}{rba}$	b = number of subsamples/sample
5) Two-Stage Sampling, Within-Lab Variation	$\sigma_{\bar{x}}^2 = \dfrac{\sigma_c^2}{c} + \dfrac{\sigma_s^2}{cs} + \dfrac{\sigma_a^2}{csa}$	r = number of composite samples

Definitions of Terms (continued):
- a = number of analyses per sample (or subsample)
- e = number of laboratories getting samples

Figure 1. Sampling models for estimating a population average.

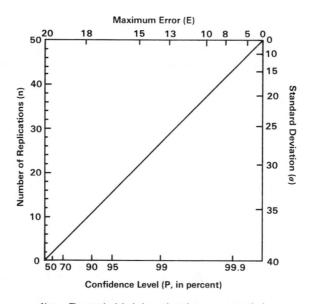

Figure 2. Nomograph to determine the number of samples (replications) required to achieve a specified maximum error.

Note: The standard deviation and maximum error must be in the same units (percent, ppb, etc.)

Table II. Selection of Sample Size (Using Equation 8)

σ_c^2 = 10% = container variance component

σ_s^2 = 40% = sample variance component

σ_a^2 = 20% = analytical variance component

E ≤ 3% Z_p = 1.96 (95% probability)

Sample Size

E	Containers c	Samples/Container s	Analyses/Sample a
2.99	30	1	1
2.92	18	2	1
2.98	26	1	2
2.99	15	2	2
2.95	11	4	1
2.94	10	4	2

---

characteristic. Evaluating composite sampling procedures requires more complex models which are extensions of the ones discussed here (7).

Allocation of Sampling Resources

Resource (cost) information can be used in conjunction with the sampling models to obtain an optimum allocation of resources in an environmental sampling study.

A model has been developed for determining a cost-effective sample size (n) when estimating cost (8).

$$C = C_0 + C_1 n + C_2 \left[b^2 + \sigma^2/n \right], \qquad (9)$$

where

n = number of samples analyzed

σ = standard deviation of samples

b = analytical bias

C_0 = overhead cost

C_1 = cost per sample analyzed

C_2 = cost of estimation error.

The equation assumes that the cost of estimation error for an estimate based on n replicate samples is proportional to the mean squared error of estimation (this includes both bias and precision). It can be shown that the value of n that minimizes equation 9 is

$$n = (C_2\sigma^2/C_1)^{1/2} . \tag{10}$$

Note that bias does not affect the optimum n (since replication does not reduce bias). The major difficulty with applying this model lies in identifying the cost of estimation error, C_2.

Even if the cost of estimation error cannot be quantified as this model requires, effective allocation of resources may be possible when detailed knowledge of sources of variation is available. In this case, a replication strategy can be based on variance component and cost information. For example, consider the problem of deciding how many samples to collect and how many analyses to perform on each sample Let

σ_1 = standard deviation due to sampling

σ_2 = standard deviation due to analysis

C_1 = cost/sample

C_2 = cost/analysis

n_1 = number of samples

n_2 = number of analyses/sample

Then the cost of sampling and analysis is

$$C = n_1 C_1 + n_1 n_2 C_2, \tag{11}$$

and the variance of the estimated population concentration (the average of $n_1 n_2$ analytical results) is

$$\sigma_{\bar{x}}^2 = \sigma_1^2 n_1 + \sigma_2^2/n_1 n_2. \tag{12}$$

Suppose we need to estimate the population concentration within $\pm$E with confidence P. The most economical allocation of extractions and analyses to meet this requirement is (3).

$$n_2 = (C_1\sigma_2^2/C_2\sigma_1^2)^{1/2} \tag{13}$$

and

$$n_1 = z_p^2(\sigma_1^2 + \sigma_2^2/n_2)/E^2. \tag{14}$$

For example, if C_1 = $48, C_2 = $20, σ_1 = 10 ppb, σ_2 = 15 ppb, P = 95% and E = 20 ppb, then $n_1 = n_2 = 2$. That is, if two samples are selected and two analyses are run on each sample, the average of the four analyses will have a maximum error of $\pm$20 ppb with 95% confidence. The cost per study will be $176.

If the maximum allowable cost is fixed, then the best n_2 is still determined by equation 13, but n is based on the cost constraint

$$n_1 \leqslant C/(C_1 + n_2 C_2) \tag{15}$$

If the maximum allowable cost in the above example is $90, then n_2 = 2 and n_1 = 1. Thus one sample and two analyses would be done at a cost of $88; the maximum probable error would be ± 29 ppb with 95% confidence.

Figure 3 is a nomograph for determining n through equation 13. The values of C_2/C_1 and σ_2/σ_1 are computed, then the line through these values gives n_2 on the middle scale. After obtaining n_2, n_1 can be obtained by the previous nomograph (Figure 2) with $\sigma^2 = \sigma^2 + \sigma_2^2/n_2$. To use the nomograph for the above example, read n_2 = 2 where the line through C_2/C_1 = 0.4 and σ_2/σ_1 = 1.5 crosses the middle scale in Figure 3. With n = 2, $\sigma = (10^2 + 15^2/2)^{1/2} = 14.6$. Then n_1 = 2 can be found using Figure 2 with a maximum error of 20 ppb and confidence level 95%.

Models such as those given in this section can be extended to more than two stages of sampling (8).

A Strategy for Designing Environmental Sampling Plans

A sample size can be determined and efficient allocation of resources accomplished when the following information is available to the study planner for each characteristic of interest:

(1) sampling objective,
(2) required precision (E),
(3) model relating E to the sample size for the important sources of variability,
(4) estimates of variance components in the model, and
(5) estimates of cost.

The administrators or users of the study results must supply the objectives and required precision. Statisticians can develop the models for alternative sampling strategies. The estimates of variance components and costs can come from a number of places:

(1) previous environmental sampling studies of similar nature (similar in the characteristics to be measured and the sampling media),
(2) research studies on analytical methods and sampling procedures,
(3) pilot studies conducted specifically to estimate unknown components of variation and costs, or
(4) theoretical considerations and experience (4).

A slight modification of the pilot study is a double-sampling plan for estimation (9,10). Double-sampling plans were developed to provide estimates with a fixed precision using as few observations as possible. If the sources of variation and variance components are known prior to the study, then a fixed sample size plan is the best

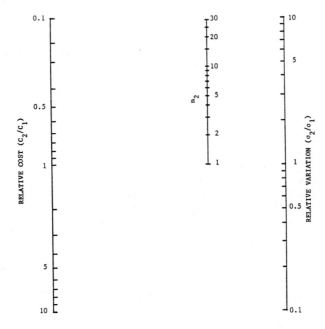

Figure 3.   Nomograph to determine the optimum number of replications in the second stage of a two-stage procedure.

design to meet the objective. If these quantities are not known, then information from the study can be used to estimate the components.

In a double-sampling plan, a small initial sample (n_4) is selected. Variance components are estimated from the results of these samples and then these estimates are used to estimate the total sample required (n). An additional $n-n_4$ samples are then collected to complete the study.

The size of the initial sample is an important consideration in a double-sampling plan. The initial sample should be as large as practicable with the constraint that only a small chance exists that n_4 is greater than n. The expected precision of estimates from the initial sample can be used as a guide in determining n_4. When the normal distribution is a reasonable model for the measurements, the t-distribution is applicable in developing confidence intervals if only estimates of variance components are available. Estimates of averages of characteristics from the n_1 sample will have confidence intervals:

$$\overline{X} \pm t_p \ S/\sqrt{n_1} \qquad (16)$$

where

$\overline{X}$ = average of n_1 measurements,
S = standard deviation of n_1 measurements, and
t_p = t-statistic, for a p% confidence level.

Estimates of the standard deviation of the n_1 measurements will have a confidence interval:

$$\text{Upper Limit} = \sqrt{\frac{V}{\chi_p^2}} \ S \qquad (17)$$

$$\text{Lower Limit} = \sqrt{\frac{V}{\chi^2_{(1-p)}}} \ S \qquad (18)$$

where V is the degrees of freedom associated with the estimate S and χ_p^2 is the tabled Chi-square statistic at p percent. Expressions similar to equations 17 and 18 can be used to obtain confidence intervals for variance component estimates in multi-stage sampling and in non-normal distribution cases.

Figure 4 shows values of the t-statistic for 90%, 95%, and 99% probability levels and for degrees of freedom ranging from two to 30. Figure 5 shows the distribution of $\sqrt{V/\chi_p^2}$ at values of χ_p^2 used to obtain a 95% confidence interval. Figures 4 and 5 can be used to evaluate initial sample sizes in a double-sampling plan. For 90 and 95 percent confidence intervals, the t-statistic becomes relatively stable after 6 to 10 degrees of freedom (n = 7 to 11). The confidence limits for the standard deviation become relatively stable after 10 to 15 degrees of freedom for the estimate. Thus, an initial sample size that will provide 8 to 15 degrees of freedom will provide reasonable

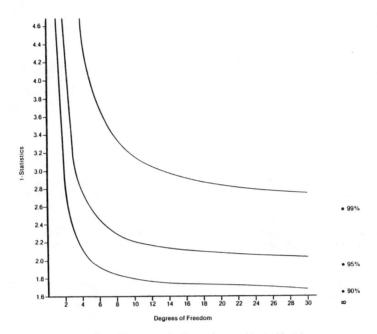

Figure 4. Nomograph for the t-distribution.

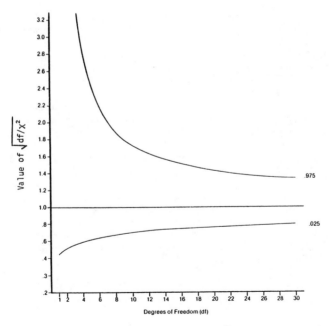

Figure 5. Nomograph for the X^2 distribution.

estimates of the parameters required for determination of the total sample size, n.

For single-stage sampling, an initial sample size of 6-15 is reasonable. For multi-stage sampling, analytical variance components, and other complications, experimental design techniques are required to develop a sampling strategy for the initial sample (11). The design of the first stage sampling effort should be such that at least 10 degrees of freedom are available to estimate each important variance component.

An Example of the Double-Sampling Strategy

The example given here illustrates the application of the double-sampling strategy discussed previously. Hydrocarbon emission rates at waste sites are estimated using a Flux Chamber. The objective is to provide an overall emission rate for the total site area. The sampling approach is to divide the waste sites into homogeneous zones (strata) and then prepare an imaginary grid for each zone. Then grid units are randomly selected for testing using a random number table.

Each zone is divided by an imaginary grid with units of approximately 4 meter^2. A series of consecutive numbers is then assigned to the units of each grid. Through the use of a random numbers table, n_K (for the Kth zone) of the grid units for sampling locations are selected, where

$$n_K \geqslant 6 + 0.1 \sqrt{\text{area of zone K } (m^2)}$$

If n_K is greater than the total number of grid units within a zone, then at least one sample should be taken within each grid unit. A minimum of six samples is required within each zone.

Calculate the emission rate of each species of interest for the n_K samples. Then compute preliminary estimates of the sample mean ($\overline{E}_K$), variance (S_k^2), and the coefficient of variation (CV_K) for each zone K (for each species).

Using Table III and CV_K, the total number of samples (N_K) to be collected from a given zone is obtained. An individual value of N_K is determined for each species of concern in each zone. The appropriate number of samples to be taken from a given zone is the largest of the individual N_K values. If the total sample size required (N_K) for a given zone is greater than the preliminary sample size (n_K), then N_K-n_K additional samples must be collected for zone K. As before, the grid units selected for additional sampling locations should be chosen at random from previously unsampled locations.

The estimates of the sample mean ($\overline{E}_K$) and variance (S_K^2) for each zone K can be calculated based on the N_K measurements. The overall sample mean ($\overline{E}$) for the total site area for each species of concern can then be calculated using

$$\overline{E} = \sum_{K=1}^{\gamma} W_K \overline{E}_K$$

where W_K is the fraction of site area represented by zone K.

Table III. Total Sample Size Required Based on the Preliminary
Sample Coefficient of Variation Estimate*

Coefficient of Variation - CV (%)**	Sample Size Required
0 - 19.1	6
19.2 - 21.6	7
21.7 - 24.0	8
24.1 - 26.0	9
26.1 - 28.0	10
28.1 - 29.7	11
29.8 - 31.5	12
31.6 - 33.1	13
33.2 - 34.6	14
34.7 - 36.2	15
36.3 - 37.6	16
37.7 - 38.9	17
39.0 - 40.2	18
40.3 - 41.5	19
41.6 - 42.8	20
42.9 - 43.9	21
44.0 - 45.1	22
45.2 - 46.2	23
46.3 - 47.3	24
47.4 - 48.4	25
48.5 - 49.5	26
49.6 - 50.7	27
50.8 - 51.6	28
51.7 - 52.3	29
52.4 - 53.4	30

*Value given is the sample size required to estimate the average
emission rate with 95% confidence that the estimate will be within
20% of the true mean.
**For CVs greater than 53.4, the sample size required is greater or
equal to $CV^2/100$.

The 95% confidence interval for each species of interest for
each zone (CI_K) is calculated using the following equation

$$CI_K = \overline{E}_K \pm t_{0.05} \sqrt{S_K^2/N_K}$$
(19)

where $t_{0.05}$ is obtained from the t-distribution for appropriate
degrees of freedom, N_K-1.
    The sampling strategy outlined is designed to provide 95% con-
fidence that the average emission rate estimate (for a given zone)
will be within 20% of the true mean.

96   ENVIRONMENTAL SAMPLING FOR HAZARDOUS WASTES

Literature Cited

1. Deming, W. E. "The Logic of Evaluation"; Handbook of Evaluation; Research; Sage Publications, Inc.:   Beverly Hills, CA, 1975.
2. "Standard Recommended Practice for Sampling Industrial Chemicals"; E300-73, American Society for Testing and Materials: Philadelphia, PA, 1979.
3. Cochran, W. G. "Sampling Techniques";   John Wiley & Sons, 1977.
4. Knotochvil, B. G.; Taylor, J. K. Analytical Chemistry July 1981, Vol. 53, No. 8.
5. Nelson, C. R. "Applied Time Series Analysis"; Holden-Day: San Francisco, CA, 1973.
6. "NSPS for SO$_2$ Emissions from Industrial Boilers," Radian Corporation, EPA 68-02-3058, Office of Air Planning and Standards, RTP NC, 1981.
7. Elder, R. S.; Thompson, W. O.; Myers, Technometrics, 1980, 22(2), 179.
8. Bennett, C. A. Franklin, N. D. "Statistical Analysis in Chemistry and the Chemical Industry"; Wiley: New York, 1954.
9. Cox, D. R. Biometrika, 1952, 39, 217.
10. Paulson, E. The Annals of Mathematical Statistics 1969, 40, 509.
11. Youden, W. J., A Public Health Service Symposium, No. 999-AP-15, 1964, p. 35.

RECEIVED August 16, 1984

Soil Sampling Quality Assurance and the Importance of an Exploratory Study

DELBERT S. BARTH[1] and BENJAMIN J. MASON[2]

[1]University of Nevada—Las Vegas, Las Vegas, NV 89114
[2]ETHURA, Derwood, MD 20855

Data from monitoring programs cannot be evaluated with confidence unless adequate quality assurance has been incorporated into the program designs. A good quality assurance program enables sources of error associated with each step of the monitoring efforts to be identified and quantified.

The component-of-variance analysis is based upon the premise that the total variance for a particular population of samples is composed of the variance from each of the identified sources of error plus an error term which is the sample-to-sample variance. The total population variance is usually unknown; therefore, it must be estimated from a set of samples collected from the population. The total variance of this set of samples is estimated from the summation of the sum of squares (SS) for each of the identified components of variance plus a residual error or error SS. For example:

$$SS_t = SS_s + SS_p + SS_{ex} + SS_a + SS_{er}$$

where SS_t = total SS
SS_s = sampling SS
SS_p = sample preparation SS
SS_{ex} = extraction SS
SS_a = analysis SS
SS_{er} = error SS

The result of this analysis provides a measure of the precision of the estimate of the mean plus confidence limits for the estimate.

Until recently, emphasis on quality assurance in support of monitoring programs has been placed on the instrumental analytical procedures. This level of quality assurance is inadequate when the medium being sampled is not homogeneous, which is particularly true for soil and sometimes may be true for air, water, sediments, or foods. For example, two soil samples taken a few feet apart may differ in chemical pollutant concentrations by an order of magnitude or more while the analytical errors may account for a negligibly small portion of the total variance. Thus, for soil monitoring programs a more comprehensive quality assurance program is mandatory.

0097-6156/84/0267-0097$06.00/0

Clearly it is not possible to separate the required quality assurance procedures for soil monitoring from the objectives of the study. Examples of objectives are:

- To determine the levels of contaminants and their spatial and temporal distribution.
- To determine the source, transport path, or receptor for a pollutant.
- To determine the presence of known or unknown contaminants in comparison to their presence in an appropriate background area.
- To provide input into risk assessments.
- To measure the effectiveness of control actions.
- To assist in a model validation study.

Administrative or legal actions may be taken on the basis of an evaluation and interpretation of monitoring data. The consequences of taking or not taking action must be understood before an allowable confidence interval can be set for the data. Often a value judgement is required concerning the acceptable probability of making a Type I (false positive) or a Type II (false negative) error. It is not possible to design a meaningful quality assurance program until this step has been taken. The Type I and Type II error for the QA/QC effort should be equal to the error levels chosen for the sampling effort itself and may range, for example, from 20% to 1% or less.

The Type I error is the error most often cited in the literature. In environmental monitoring, however, the Type II error may be more important. A false negative could create major problems for the environmental manager if it suggests that a cleanup is not necessary when in fact action levels are being exceeded.

There may be a temptation to avoid making the necessary value judgments concerning acceptable probabilities of making different kinds of error. Instead, it may be easier to adopt the guiding principle that one should always strive to achieve the highest precision and level of confidence (or lowest error) possible with existing available resources. Such an approach will rarely be cost-effective. Two consequences are possible. The data may be much better than required which indicates resources have been wasted, or the data may not be of adequate quality thereby resulting in costly decisions which may be wrong. Resource availability is always an important factor for consideration in the establishment of quality assurance programs, but resource availability should not be the sole determinant of required quality assurance methods and procedures.

Establishing QA Objectives

The steps outlined below are intended to guide the development of data quality objectives for the sampling effort. These have been discussed in part by others (1,2).

1. Identify the objectives of the study. These should reflect the specific items of information needed to make decisions following completion of the study.

2. Determine the components of variance that should be built into the statistical design. Proper stratification of the study area will allow identification and quantification of several sources of variation. The sources of variation that can be controlled by the sampling are determined by the particular sampling design and by the pattern of sample collection superimposed over the area. An analysis of variance of the data provides estimates of the components of variance.
3. Choose the appropriate confidence level. Generally, a confidence level of 95% or better is desired. However, this is often not possible because of economic or other constraints. The investigator may have to recalculate the confidence level that can be reasonably attained with the resources available. If that revised level is not adequate to allow achievement of the study objetives, more resources must be found, the study objectives revised, or the study cancelled. The major point is that the confidence level should be explicitly recognized before the study is conducted and not after the data are collected.
4. Obtain sampling data from other studies that have similar characteristics to the one being designed. Especially useful are results obtained from replicated samples. These data can be used to derive estimates of data quality indicators in the early stages of the sampling process.
5. Calculate the mean and range of each set of replicates.
6. Group the sets of replicates according to concentration ranges and by the types of samples that are believed to be similar. An example of the groupings might be samples in the range from 0 to less than 10 ppm, 10 to less than 25, etc. Another grouping might be by soil type such as sand, silt, or clay.
7. Calculate the critical difference $[R_c]$. For any group of duplicate analyses that are considered similar to each other, their ranges $[R_i]$ and means $[\overline{X}_i]$ can be used to estimate R_c. A similar R_c would be expected for future duplicate analyses at similar concentration levels $[C]$.

$$R_c = \frac{3.27 \ [C]}{n} \left[\sum_{i=1}^{n} \frac{R_i}{\overline{X}_i} \right]$$

where $\overline{X}_i = \dfrac{x_i + x_{i+1}}{2}$ and $R_i = x_i - x_{i+1}$

8. Develop a table of R_c values for various concentrations that span the range of concentrations of interest. (A similar approach makes use of confidence limits based on the standard deviation rather than the range.)

These data are used to accept or reject a set of replicated samples. The replicates are usually duplicate samples. Therefore, the difference between the two values should lie within the critical range. If not, the sample is rejected and the analyses rerun if possible. Discarding results should only be done after careful review of the data. There are situations in soil sampling where the coefficient of variation can reach hundreds of percent due to the variability in the soil system; therefore, suspected outliers may, in fact, be a part of a wide distribution. This tends to be the case when very high levels of chemicals have been spilled over small areas or where chemicals have flowed through desiccation cracks, animal burrows, or old root channels. Observations made by the field party, and noted in the log books at the time of sample collection, can aid in deciding whether or not to discard a sample.

9. Use the preliminary R_c table until data are acquired during the sampling. As the analyses proceed, the results are combined with those from previous studies. When approximately fifteen pairs (1) of results are acquired from the particular study area, a new table should be calculated based upon the average range of the data that has been accepted to date.

10. Use the data collected during the preliminary or exploratory site investigation as the data base for designing later studies.

Role of the Exploratory Study

Once objectives have been defined, a study protocol including an appropriate QA/QC program is developed. Initially, both literature and information searches should be made. If possible, selected field measurements based on an assumed dispersion model can also be made. The objective of the exploratory study is to obtain the best possible answers to the following questions.

- What are the likely sources of the pollutants of concern?
- How have these sources varied in the past compared to their present emissions?
- What are the important transport routes contributing to the soil contamination?
- What is the geographical extent of the contamination?
- What average concentrations of the pollutants exist at different locations, and how do these vary as a function of location and time?
- Do localized areas of high concentrations exist and, if so, where are they and what are their concentrations?
- Is it possible to stratify the sampling region in such a way as to reduce the spatial variations within strata?
- What are the soil characteristics, hydrogeological factors, meteorological or climatic factors, land use patterns, and agricultural practices affecting the transport and distribution of the pollutants of concern in soil?
- What is an appropriate background, or control region, for the study region?
- What are the acceptable levels of precision for both Type I and Type II errors for this study?

Of course, if detailed and specific answers to all these questions were available in advance, there would be no need to conduct the study.

When dealing with an emergency situation such as a spill of a hazardous chemical, there usually is not time to proceed in the

deliberate fashion contemplated here. Instead, it may be necessary
to compress the planning of the study into a very short time period
and proceed to the final definitive study without delay. In the
following discussion, however, we will assume that a reasonable
amount of time is available to conduct an exploratory study prior to
the more definitive study.

Much information pertinent to the above questions may already
be available in the published literature, in the files of Government or industry, in research reports of local universities, or in
the knowledge of local citizens. A carefully planned effort should
accumulate relevant information that is available. Usually, a fixed
period of time should be allowed for the collection and analysis of
this information. Then the design and implementation of the field
measurements portion of the exploratory study should be undertaken.

It is not possible to separate the QA/QC from the total soil
monitoring study design. As previously noted, the objectives of the
monitoring study are the driving force for all elements of the
study design including the QA/QC aspects. The end product of the
exploratory study is information and field data that will serve as
the basis for the design of a definitive monitoring study including
an integrated QA/QC program. One element of this total QA/QC program
is the soil sampling QA/QC plan.

Number and Locations of Sites for Sampling

The definitive study design should designate the appropriate number
of sampling sites at the appropriate locations so that determinations
of mean concentrations and standard deviations for the regions of
interest may be made. The number of sites required in a given region
can be calculated if one knows the required precision, the standard
deviation of the mean, and the required levels of confidence (related
to acceptable levels of Type I and Type II errors). These must be
designated by the party that will use the results of the soil sampling
study. The standard deviation of the mean of the total population of
soil samples in the study region must be estimated on the basis of
the standard deviation of a suitable sample taken from the total
population during the exploratory study.

The locations where soil samples are taken can be selected on
the basis of random, judgmental, or systematic sampling designs. A
major input into selecting the optimum sampling design will be the
information accumulated prior to the field sampling phase. Usually
the optimum approach will be a combination of judgmental and systematic or random sampling. A model may be hypothesized describing a
likely spatial distribution of soil contamination as well as a likely
control area. Selection of the number and location of sampling sites
on the basis of such a model is a judgmental approach.

As an example, suppose it is suspected that an abandoned hazardous waste site is leaking wastes into the ground water. Further,
suppose that the ground water is being used to irrigate crops in the
vicinity. Preliminary information has identified some of the pollutants that have been placed in the waste site and has determined the
hydraulic gradient extending from the site location. The recommended
approach is to establish a radial grid system with the center at the
waste site and the zero azimuth line along the direction of the

hydraulic gradient. The largest number of samples would then be taken along the zero azimuth and along the $\pm 45°$ azimuth from zero. This is judgmental sampling. However, to make sure that important data are not missed some additional samples should be taken close to the waste area along every 45° azimuth (5 additional directions). This adds systematic sampling to take care of cases where, for example, some immiscible waste constituents may be moving in a direction different from the hydraulic gradient, the hydraulic gradient has not been properly defined, or there are other sources contributing to the soil contamination. The location of the samples taken along each axis and in the pollutant plume would be located at random.

For the selection and sampling of a control area, a combination of judgmental and random sampling is recommended. Based on the available information and the assumed transport model, select a background or control region which is like the study area in every important particular except for the absence of contamination. Select 6-15 sampling locations at random from the background area to obtain data for calculating the mean and standard deviation of the concentrations of selected waste constituents.

The QA/QC program for the exploratory study need not be as stringent as that for the more definitive study. Keep in mind, however, that reasonable levels of precision and confidence must be attained for the resulting data to serve as an adequate foundation for further studies. As a minimum, it is suggested that duplicate samples be collected from at least 5% of all sampling locations and that at least 5% of all samples be split into triplicate samples. Furthermore it is recommended that at least 12-15 additional independent QA/QC soil samples be taken on a random basis at approximate midpoints between selected sampling points in regions where the hypothetical model predicts the highest concentrations will be found.

Duplicate sample results will help to establish precision among different samples collected from the same site. Triplicate splits of samples will give a measure of precision within a single sample which tests the homogeneity of the sample. The additional 12-15 QA/QC samples will provide data to use in evaluating possible changes in means and standard deviations when additional sampling points are added. If the two groups of samples are equal, the samples can be combined. If not, then there is an indication of some form of bias in the sampling design. This bias must be carefully evaluated in order to determine if additional sampling is needed. In addition to the above QA/QC checks on sampling, all normal analytical QA/QC procedures should be operative for the exploratory study.

Sampling and Sample Handling

A protocol must be established and followed for sample preparation, labeling, packaging, shipping, and chain-of-custody procedures. Also, the volume of the samples will be specified by the analytical laboratory depending on the analytical methods to be used and the desired sensitivity. Accordingly, principal attention will be given here to the sampling methods, preparation of the samples for analysis, and QA/QC aspects of both.

Some major concerns in sampling include required depth of sampling, whether or not sequential samples at different depths will be

required, whether or not samples should be composited, frequency of
sampling, and sample preparation for analyses, as well as the QA/QC
aspects of all these concerns. In deciding how to deal with these
concerns, the objectives for which the soil monitoring study is being
conducted must be kept in mind. The exploratory study provides a
limited opportunity to investigate some of the above areas experimentally in order to determine what effect the sampling parameters
may have on the QA/QC aspects of the total study. The expenditure of
modest additional resources in the exploratory study may well lead to
more cost-effective designs for the final definitive study.

The simplest adequate sampling device should be used. Where the
contaminant is believed to be on the surface, a soil punch or trowel
may be used. If the contaminant is soluble or is expected to be
located more than a meter below the surface, a truck mounted core
sampler such as a split spoon sampler should be used.

Surface sampling should be augmented with 12-15 sequential
samples taken down to 1.5 meters in order to determine if the
pollutant has moved downward. These 12-15 additional samples should
be located in the area of major contamination.

With regard to compositing of samples, the major concerns are
that the sample be representative and that high concentrations from
limited areas not be significantly reduced by being averaged with
lower level samples. It is recommended that at least four different
samples taken in the vicinity of each selected sampling site be
composited into a single sample. A single sample which is not composited should be collected for comparison with the composited
samples.

The exploratory study is not designed to obtain information on
temporal patterns in the chemical concentrations since these studies
are expected to be completed in a short period of time. If it is
possible to select the time for the exploratory study, it should be
conducted at a time when the concentrations would be expected to be
at a maximum. It may be necessary to use the hypothesized dispersion
model in order to make this decision. For example, the sampling
normally should not be done immediately following a heavy rain, when
the ground is frozen solid, or when a wind is blowing at 20 to 30
knots. Temporal trends will have to be addressed in the final study.

Sample preparation for analyses introduces some possibilities for
errors. Vegetation, sod, or other non-soil material must be removed
from the sample. This is followed by grinding or mixing the sample
in some way, sieving it, and then drying it when necessary.

The grinding and mixing devices, as well as any sieves, must
be carefully cleaned between each sample in order to avoid cross-
contamination of the samples. The final rinse water should be sampled
on 5% of the decontamination cycles in order to provide a blank for
use in evaluating the decontamination efficiency. These samples
should be submitted to the laboratory along with the other QA/QC
samples.

Perhaps the most serious possibility for error at this stage of
the sampling process is in discarding of vegetation, sod, or other non-
soil material collected along with the soil sample as well as the
discarding of other materials retained on the sieve. It is recommended that for approximately 10% of all samples where vegetation,
sod, or other non-soil material is discarded, all discarded material

be retained (including the materials retained on the sieve) and sent to the analytical laboratory for bulk analysis. The results of these analyses will give a quantitative estimate of the possible errors introduced by eliminating these non-soil materials.

Analysis and Interpretation of Data

Analysis and interpretation of the information and data resulting from the exploratory study will provide the basis for designing the final definitive monitoring study including all elements of the QA/QC plan. For example, decisions must be made on whether or not the selected control area is adequate and appropriate; whether the hypothesized model is valid; whether the study area should be stratified and if so, how; what number of samples should be collected at what locations; and whether or not the QA/QC plan for sampling is adequate and if not, how it should be changed.

If the exploratory study is conducted well, it will provide some data for achieving the overall objectives of the total monitoring study. It will provide a check of the feasibility and efficacy of all aspects of the monitoring design including the QA/QC plan. It will serve as a training vehicle for participants. It will pinpoint where additional measurements need to be made. Finally, it will provide a body of information and data which may be incorporated into the final report for the total monitoring study.

Literature Cited

1. "Guidelines Establishing Test Procedures for the Analysis of Pollutants; Proposed Regulations," U.S. Environmental Protection Agency, Federal Register, 44, pp. 69464-69575, 1979.
2. Bauer, Edward L. "A Statistical Manual for Chemists"; Academic Press: New York, New York, 1971.

RECEIVED August 6, 1984

Quality Assurance for a Measurement Program

JOHN K. TAYLOR

National Bureau of Standards, Washington, DC 20234

Measurement data of adequate quality are necessary if environmental problems are to be properly understood and effective actions taken to correct them. Quality assurance programs for all aspects of the measurement process are necessary to provide such data. In previous papers, the general aspects of quality assurance of measurements (1,2) and the essential features of a laboratory quality assurance program (3) have been presented. This paper discusses the general features of quality assurance for measurement programs, and especially monitoring programs.

In a classical paper, Deming presented fourteen points for management of quality and productivity (4). Three of these points are particularly relevant for monitoring programs and may be paraphrased as follows:

- The new philosophy: we can no longer accept defects.
- Depend on vendors of services that use process control.
- Price has no meaning without evidence of quality. Demand statistical control by vendors with supporting evidence.

Monitoring programs involve the participation of vendors of services in a hierarchical array.

- Monitoring program managers
- Laboratory managers
- Individual scientists

Following the Deming philosophy, each level depends on those below it in a vendor relationship. Thus, quality assurance is effective only if each level performs in an effective manner.

Quality Assurance Responsibilities

The individual scientist is the key to the production of quality data and must have technical competence and a dedication to quality work. Passive following of good laboratory practices and good measurement practices is not enough. Involvement in their development is required if the quality assurance program is to be credible. Standard operations procedures (SOPs) do not minimize the need for technical competence. Analysts exhibit varying degrees of proficiency when

using the same (SOPs) and similar equipment. Skill and judgment are
still required if sound data are to result.
The laboratory has the resposibility for establishing and main-
taining an effective quality assurance program. It must provide
adequate facilities and equipment, employ a competent staff, and
create an atmosphere that encourages excellence. It must have
quality as its prime goal and must maintain adequate supervision
of the measurement process, including the release of data (3). In
this regard, it must operate its own quality assessment program
using reference materials (5), internal quality assessment samples,
and control charts (6) to verify the quality of its data output.
Data produced by different individuals within a laboratory need
to be coordinated. This is a further responsibility of laboratory
management.
Monitoring programs must have their own quality assurance programs. These may be called project quality assurance plans or protocols for specific purposes (3). If reliable vendors of services are
used, the bulk of the quality assurance effort can be placed on those
activities unique to the program. Without reliable vendors, QA
efforts will be ineffective since it is not cost effective to police
quality assurance practices at all lower levels nor to screen all
data for its validity.

Monitoring Program Responsibilities

A prime responsibility of the overall program management is determination of program objectives. Data quality objectives must be set and
reliable vendor laboratories selected, evaluated for their initial
qualifications, and audited for their ongoing proficiency. The
monitoring program must also provide for an intercalibration program
which involves the vendor laboratories. Reliable vendors will maintain their own quality assurance programs and provide evidence of the
quality of their data. The program QA plan will specify the control
charts and other records that should be maintained by the vendors and
used in judging the quality of the data output.
The program must require the vendors to measure a number of
reference samples and/or duplicates submitted in a planned sequence.
It should require prompt measurement and reporting of these data and
should maintain the results in a control chart format. Prompt feed-
back and follow-up of any apparent data discrepancies and reconcilia-
tion of the results with control charts maintained by the vendors
are required to minimize the length of uncertain performance. The
quality assurance plan should include random sampling of the vendors'
data for their validity and conformance with quality assurance requirements. If quality assurance is properly practiced at all levels,
an inspection of 5 percent of the total data output should be
adequate.
System audits of vendors' quality assurance programs should be
made initially and at random intervals throughout the duration of a
monitoring program. The general procedure described by Gaft and
Richards is recommended for this purpose (7). The monitoring program plan should specify that vendors conduct internal systems audits
similar to the above, but at more frequent intervals. Records of

these should be available for inspection on request. Such a procedure would minimize the number of system audits that need to be conducted.

Since the use of competent vendors is the key to a successful monitoring program, a system for their pre-qualification is necessary. Only laboratories experienced and skilled in the methodology to be used should be considered. Candidate laboratories must present evidence of proficiency equal to or exeeding the minimum requirements of the program. Retention of vendor relationships and payment for services should be contingent on maintenance of proficiency standards, specified in advance. Faulty data are unworthy of payment. Program management will need to be fair in its judgment of such matters and may need to establish arbitration procedures in this regard.

The program quality assurance plan may need periodic or emergency revisions. Ongoing review of the data should reveal whether any deficiencies are due to inadequate performance of vendors or to defects in the quality assurance plan. Defects in the plan could result from inadequate quality assessment techniques if measured levels of contaminants were significantly different from anticipated levels, for example.

Reference Laboratory

The monitoring program may find it advantageous to use a reference laboratory to carry out some of the responsibilities outlined above. A reference laboratory may be used for arbitration and as a third-party participant to provide smooth operations of the program. A reference laboratory should have unquestioned competence and demonstrated experience related to all aspects of the monitoring program. Above all, it should be recognized for its objectivity in scientific investigations, must have the facilities and equipment to conduct reliable referee measurements, and should have the capability for real-time responses to questions as they arise. To the extent that it might be delegated overall responsibilities, it should have program management experience, as well. Obviously, it must have its own impeccable quality assurance program which can serve as a pilot for the monitoring program and as a model for the vendors.

The duties of the reference laboratory might include the following typical assignments:
- o Select methodology
- o Pre-test methodology
- o Develop reference materials
- o Develop protocols
- o Pre-test protocols
- o Set performance standards
- o Qualify candidate laboratories
- o Conduct performance and systems audits
- o Ongoing data appraisal
- o Troubleshooting of problems
- o Data release decisions
- o Training of vendors

Conclusion

Monitoring operations have considerable analogy to modern manufacturing operations involving a number of subcontractors or vendors of
services. The monitoring program should establish its own quality
standards and specify minimum quality assurance practices that the
vendor of services should follow. The monitoring program should also
supply reference materials for intercalibration of the vendors and
maintain control charts to monitor intercalibration and to detect any
deviations from proficient performance. Only experienced vendor
laboratories, pre-qualified by objective testing procedures, should
be used. Their maintenance of proficiency should be monitored
throughout the program and their retention as vendors should be contingent on their ability to supply data of specified quality. Laboratories should not be paid for data of insufficient or undocumented
quality.

On occasion, an adequate number of qualified vendors may not be
available. In these cases, the monitoring program management must
select and train the most likely candidates prior to their accep-
tance. In no case should monitoring programs be initiated with
unqualified vendors and pressures to do so should be strongly
resisted.

Literature Cited

1. Taylor, J. K. Anal. Chem. 1981, 53, 1588A-96A.
2. Taylor, J. K.   "In Quality Assurance for Environmental Measurements"; ASTM STP 867: Philadelphia, PA, 1984.
3. Taylor, J. K. In "Statistics in Environmental Science"; ASTM STP
   867: Philadelphia, PA.
4. Deming, W. Edwards. "Quality Productivity and Competitive Posi-
   tions"; MIT Press: Cambridge, MA.
5. Taylor, J. K. J. Testing and Evaluation 1983, 11, 355-7.
6. "Presentation of Data and Control Analysis", ASTM STP 15D:
   Philadelpia, PA, 1976.
7. Graft, S.; Richards, F. D.   In "Evaluation and Accreditation of
   Inspection and Test Activities"; ASTM STP 814: Philadelphia,
   PA, 1984.

RECEIVED August 22, 1984

New Ways of Assessing Spatial Distributions of Pollutants

ANDRE G. JOURNEL

Stanford University, Stanford, CA 94305

In the last few years, detection, characterization, and handling of
pollution sites have grown into a new discipline with its specific
technology, initially drawn from other disciplines such as statistics,
then customized to meet the specificity of pollution control.
   A body of specific techniques already exists for sample design
and data capturing. However, the algorithms used to interpret these
data and map them are still very classical. Passkey interpolation
algorithms are most often used even though they ignore the pattern of
spatial distribution and correlation particular to each pollution
plume. Arbitrary risks α and β are used to make critical decisions,
as if agricultural standards, for example, are also valid when dealing with health hazards.
   Geostatistical techniques, such as variography and kriging, have
been recently introduced into the environmental sciences ($\underline{1}$). Although kriging allows mapping of the pollution plume with qualifica-
tion of the estimation variance, it falls short of providing a truly
risk-qualified estimate of the spatial distribution of pollutants.
   Ideally, to characterize the spatial distribution of pollution,
one would like to know at each location $\underline{x}$ within the site the
probability distribution of the unknown concentration p($\underline{x}$). These
distributions need to be conditional to the surrounding available
information in terms of density, data configuration, and data values.
Most traditional estimation techniques, including ordinary kriging,
do not provide such probability distributions or "likelihood" of the
unknown values p($\underline{x}$). Utilization of these likelihood functions
towards assessment of the spatial distribution of pollutants is
presented first; then a non-parametric method for deriving these
likelihood functions is proposed.

The Process of Estimation

Pollutant concentrations, just like metal grades in a mineral deposit,
can be seen as variables which are functions of their spatial coordi-
nates $\underline{x}$ = (u,v) or $\underline{x}$ = (u,v,w). Consider a particular pollutant
p and its concentration value p($\underline{x}$) at location $\underline{x}$.

0097-6156/84/0267-0109$06.00/0

It is usual to visualize the spatial distribution of $p(\underline{x})$ within an area A by contouring measured or estimated values $p^*(\underline{x}_k)$ at the nodes $\underline{x}_k$ of a grid. The resulting estimated map necessarily differs from the "true" map that would have been obtained from the true values $p(\underline{x}_k)$. Thus, there appears a need to characterize the potential departure $p(\underline{x}_k)-p^*(\underline{x}_k)$ between each estimate and the corresponding true value. Furthermore, there is a need to characterize the joint departure of all estimates from all true values. Indeed, a map locally precise (for each $\underline{x}_k$) is not necessarily the most precise for a global criterion such as the reproduction of spatial trends.

Probabilistic techniques of estimation provide some insights into the potential error of estimation. In the case of kriging, the variable $p(\underline{x})$ spread over the site A is first elevated to the status of a random function $P(\underline{x})$. An estimator $P^*(\underline{x})$ is then built to minimize the estimation variance $E\{[P(\underline{x})-P^*(\underline{x})]^2\}$, defined as the expected squared error (2). The kriging process not only provides the estimated values $p(\underline{x}_k)$ from which a "kriged" map can be produced, but also the corresponding minimum estimation variances $\sigma_K^2(\underline{x}_k) = E\{[P(\underline{x}_k)-P^*(\underline{x}_k) \, (\underline{x}_k)]^2\}$. These variances $\sigma_K^2(\underline{x}_k)$ can also be contoured, thus providing a map of iso-values of reliability (see Figure 1).

The Limitation of Data-free Estimation Techniques

A great deal has been written about the optimality of the kriging estimates and about the error characterization provided by iso-variance maps such as that of Figure 1. However, kriging, and more generally data-free estimation algorithms, have several drawbacks.

1. Firstly, the kriging estimator is optimal only for the least square criterion. Other criteria are known which yield no more complicated estimators such as the minimization of the mean absolute deviation (mAD), $E\{|P(\underline{x})-P^*(\underline{x})|\}$, yielding median-type regression estimates.

2. Secondly, knowledge of the estimation variance $E\{[P(\underline{x})-P^*(\underline{x})]^2\}$ falls short of providing the confidence interval attached to the estimate $p^*(\underline{x})$. Assuming a normal distribution of error in the presence of an initially heavily skewed distribution of data with strong spatial correlation is not a viable answer. In the absence of a distribution of error, the estimation or "kriging" variance $\sigma_K^2(\underline{x})$ provides but a relative assessment of error: the error at location $\underline{x}$ is likely to be greater than that at location $\underline{x}'$ if $\sigma_K^2(\underline{x})>\sigma_K^2(\underline{x}')$. Iso-variance maps such as that of Figure 1 tend to only mimic data-position maps with bull's-eyes around data locations.

3. Thirdly, and most importantly, both the kriging algorithm and kriging variance are independent of data values, thus not distinguishing the estimation of extreme values (usually the most important for pollution control) from the estimation of median or background values.

Consider, for example, a site characterized by a highly skewed distribution of pollutant concentrations, as apparent in the histogram of data values of Figure 2a. These values present a coefficient of

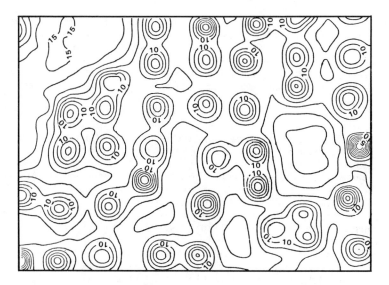

Figure 1.   Isopleths of kriging estimation variances.   (The bull's-eyes reflect the data locations.)

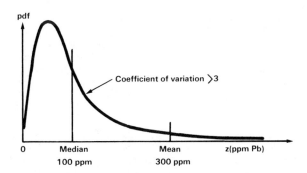

2a - Typical example of histogram of pollutant data
(note the strong positive skewness)

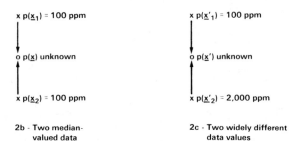

Figure 2.   The weighting algorithm should be dependent on data values.

variation (σ/m) greater than 3, which is not an uncommon figure for pollution data. Consider then the two identical data configurations of Figures 2b and 2c, where a nodal point $\underline{x}$ is estimated by two equidistant data points at locations $\underline{x}_1$ and $\underline{x}_2$ Kriging, just like more traditional estimation technique, gives an equal weight (1/2) to each of the two data and the same estimation variance value for both configurations.

Next, consider that the first configuration (Figure 2b) includes two median-type data values, say around 100 ppm Pb, whereas the second configuration (Figure 2c) includes a median-type datum value (100 ppm Pb) with an outlier datum value (2,000 ppm Pb). Clearly the potential for error attached to the second realization is greater; in other words, the estimation variance should be dependent on data values and not only dependent on data configuration. Also, there is no reason for the two weights of the second realization to be equal, that is the weights should also be dependent on data values. In some cases, e.g. nugget-type gold mineralizations, the very high grades have a smaller range of spatial correlation than background to median grades; other cases, e.g. high concentrations of pollutants, have a larger range of spatial correlation linked to the pollution plume whereas the background concentrations tend to be more erratic in their spatial distribution. In the first cases, the high datum value of Figure 2c should be downweighted; in the latter cases, it is the lower datum value that should be downweighted.

The Conditional Distribution Approach

One way to introduce the data values into the estimation algorithm is to consider the conditional probability distribution of the unknown $P(\underline{x})$, given the N data values used to estimate it. Denote this conditional cumulative distribution function (cdf) by:

$$F_{\underline{x}}(z|(N)) = \text{Prob}\{P(\underline{x}) \leq z \,|\, P(\underline{x}_1) = p_1,\ldots,P(\underline{x}_N) = p_N\} \qquad (1)$$

This conditional cdf is a function not only of the data configuration (N locations $\underline{x}_i$, i=1,...,N) but also of the N data values (p_i, i=1,...,N). Its derivative with regard to the argument z is the conditional probability density function (pdf) and is denoted by:

$$f_{\underline{x}}(z/(N)) = \frac{\partial F_{\underline{x}}(z/(N))}{\partial z} \qquad (2)$$

This conditional pdf can be seen as the likelihood function of the unknown value $p(\underline{x})$.

Moments of this conditional distribution can be written as standard Riemann integrals of the pdf $f_{\underline{x}}(z|(N))$ or as Stieltjes integrals of the cdf $F_{\underline{x}}(z|(N))$. For example, the conditional expectation is written:

$$E\{P(\underline{x})|(N)\} = \int_0^\infty z.f_{\underline{x}}(z|(N))dz = \int_0^\infty z.dF_{\underline{x}}(z|(N)), \qquad (3)$$

and is a function (usually non-linear) of the N data values.

Techniques for derivation or estimation of the cdf $F_{\underline{x}}(z|(N))$ are presented in a later section. For now, it is shown how knowledge of that cdf provides a solution to the three problems previously mentioned.

1. Firstly, various criteria for estimation, different from the least square $E\{[P(x)-P*(x)]^2\}$, may now be considered. Consider a general loss function $L(e)$, function of the error of estimation $e = p(\underline{x}) - p*(\underline{x})$. The objective is to build an estimator that would minimize the expected value of that loss function, and more precisely, its conditional expectation given the N data values and configuration.

$$E\{L(P(\underline{x})-P*(\underline{x}))|(N)\} = \int_o^\infty L(z-p*(\underline{x})).f_{\underline{x}}(z|(N))dz \qquad (4)$$

For any predetermined loss function L, this conditional expectation appears as a function of both the estimated value $p*(\underline{x})$ and the N data values, p_i, $= i=1,\ldots,N$. The optimization process consists of determining the particular value $p*(\underline{x})$ that would minimize expression (4). The solution is straightforward for some particular functions L:

-   If $L(e) = e^2$, i.e. the loss is proportional to the squared error, the least square criterion is apparent, and the best estimator $P(\underline{x})$ is the conditional expectation defined in (3). Note that this estimator is usually different from that provided by ordinary kriging for the simple fact that expression (3) is usually non-linear in the N data values.
-   If $L(e) = |e|$, i.e. the loss is proportional to the absolute value of the error, the best estimator is the conditional median, i.e. the value:

$$q_{.5}(\underline{x}) \text{ such that: } F_{\underline{x}}(q_{.5}(\underline{x})|(N)) = .5 \qquad (5)$$

Again this conditional median estimator is usually a non-linear function of the N data values.

-   If $L(e) = 0$ for $e = 0$, and infinity for any error different from zero, the best estimator is the maximum likelihood estimator, i.e. the value z_0 for which the conditional pdf $f_{\underline{x}}(z|N))$ is a maximum.

More generally, the loss function need not be symmetric: $L(e) \neq L(-e)$. Indeed, underestimation of a pollutant concentration may lead to not cleaning a hazardous area with the resulting health hazards. These health hazards may be weighted more than the costs of cleaning unduly due to an overestimation of the pollutant concentration. The optimal estimators linked to asymmetric linear loss functions are given in Journel (3).

2. Secondly, knowledge of the conditional cdf $F_{\underline{x}}(z|(N))$ provides the confidence intervals which are data values-dependent but independent of the particular estimate $p*(\underline{x})$ retained:

$$Prob\{P(\underline{x})\varepsilon]q_w(\underline{x}), \, q_{1-w}(\underline{x})]|(N)\} = 1-2w, \qquad (6)$$

with $q_w(\underline{x})$ being the conditional w-quantile such that:

$$F_x(q_w(\underline{x})|(N)) = w\varepsilon[0,1].$$

The estimate $p^*(\underline{x})$ retained need not be at the center of the confidence interval $]q_w(\underline{x}), q_{1-w}(\underline{x})]$. Since the confidence intervals are obtained directly, there is no need to calculate the estimation variance, nor to hypothesize any model for the error distribution.

3. Thirdly, and most importantly, the conditional cdf $F_x(z|(N))$ is dependent on data values accounting for the possible difference in spatial correlation between high-valued concentration data and background concentration data. This fact stems from the process of estimation of $F_x(z|(N))$ itself described in a later section.

Risks of Incorrect Decisions

The availablility of an estimate of the conditional cdf $F_x(z|(N))$ at each nodal joint $\underline{x}$ allows an assessment of the risks $\alpha(\underline{x})$ or $\beta(\underline{x})$ of making wrong decisions. Consider the contour map of a particular estimate $p^*(\underline{x})$, $\underline{x} \varepsilon A$ (see Figure 3a). Suppose that the threshold value 500 has been selected to declare any sub-area of A hazardous. The contour line 500 on Figure 3 delineates the zones which are candidates for cleaning. Within these zones, the probability that the concentration is actually under 500, i.e. the risk $\alpha(\underline{x})$ of cleaning unduly, can also be mapped:

$$\alpha(\underline{x}) = \text{Prob}\{P(\underline{x}) \le 500|(N)\} = F_x(500|(N)) \qquad (7)$$

$$x : p^*(\underline{x}) > 500$$

The probability of making a correct decision to clean is $1-\alpha(\underline{x})$, and has been mapped on Figure 3b: Most of the time it appears that this probability is less than 50 percent except in the central zone next to the pollution source. Besides changing the threshold value 500, one way to improve this probability is to take more samples (increase the number of data N) which would decrease the variance and skewness of the conditional pdf $f_x(z|(N))$.

Similarly, within the complementary zones for which the estimate $p^*(\underline{x})$ is $\le$ 500, the risk $\beta(\underline{x})$ of not cleaning unduly can be mapped:

$$\beta(\underline{x}) = \text{Prob}\{P(\underline{x}) > 500|(N)\} = 1 - F_x(500|(N)) \qquad (8)$$

$$\underline{x} : p^*(\underline{x}) \le 500$$

Estimation of the Conditional Distribution

There are essentially two approaches to estimating the conditional cdf $F_x(z|(N))$.

1. The first approach consists of assuming some multivariate distribution model for the random function $P(\underline{x})$, $\underline{x}\varepsilon A$. A convenient

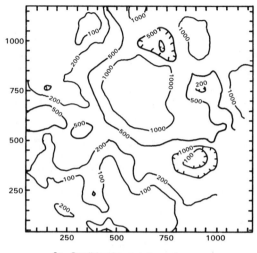

3a · Conditional expectation estimate
$p^*(\underline{x})$ expressed in ppm

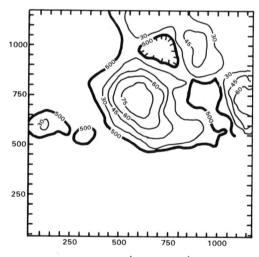

3b · $1 \cdot \alpha(\underline{x})$ $\text{Prob}\{P(\underline{x}){>}500 \mid (N)\}$,
expressed in percent (thin lines)
plotted for $\underline{x} : p^*(\underline{x}){>}500$ ppm (thick line)

Figure 3. Isopleth maps of the concentration estimate and the associated probability $(1 - \alpha(\underline{x}))$ to make a correct decision to clean.

model is the multi-ϕ-normal distribution which assumes that the normal score transform random function $U(\underline{x}) = \phi(P(\underline{x}))$ is multinormal distributed. As a consequence, the multivariate distribution of $U(\underline{x})$ and hence of $P(\underline{x})$ is fully characterized by the covariance function $C(\underline{h}) = E\{\overline{U}(\underline{x}).U(\underline{x} + \underline{h})\}$, experimentally inferred from the normal score data. Knowledge of the multivariate distribution of $P(\underline{x})$ allows exact derivation (not an estimation) of any conditional cdf $F_{\underline{x}}(z|(N))$, (4,5). The lognormal version (ϕ = Ln) of this approach is of current use in the mining industry (6). Of course the whole procedure is distribution-dependent, capitalizing on the initial multi-ϕ-normal distribution hypothesis.

2. The second approach consists of estimating sequentially the values $F_{\underline{x}}(z_k|(N))$ for a series of cut-off values z_k covering the range of variability of the concentration values, usually $[0, z_{max} < 100\%]$. The key idea is to interpret the conditional cdf $F_{\underline{x}}(z|(N))$ as the conditional expectation of an indicator transform $I(\underline{x};z)$ which can be estimated from the corresponding indicator data. Indeed, consider the indicator transform:

$$I(\underline{x};z) = \begin{cases} 1, \text{ if } P(\underline{x}) \leq z \\ \\ 0, \text{ if not} \end{cases}$$ (9)

Then:

$$E\{I(\underline{x};z)\} = 1 \times \text{Prob}\{P(\underline{x}) < z\} + 0 \times \text{Prob}\{P(\underline{x}) > z\}$$

$$= \text{Prob}\{P(\underline{x}) \leq z\} = F(z)$$

Similarly:

$$E\{I(\underline{x};z)|(N)\} = \text{Prob}\{P(\underline{x}) \leq z|(N)\} = F_{\underline{x}}(x|(N))$$ (10)

The projection theory indicates that a projection estimator of the unknown $I(\underline{x};z)$, such as a kriging estimator, is also a projection estimator (kriging) of its conditional expectation $F_{\underline{x}}(z|(N))$, (7). Consider then the indicator kriging estimator:

$$[I(\underline{x};z)]^* = \sum_{i=1}^{N} \lambda_i(z).I(\underline{x_i};z)$$ (11)

The weights $\lambda_i(z)$ are cut-off z-dependent, and are determined as solutions of a linear system (8). In fact, the indicator estimators used for the case-study underlying Figures 3 were obtained by a slightly more elaborate technique called "probability kriging or PK" (9, 10).

The kriging process (11) is repeated as many times as there are different cut-offs (z_k) retained to discretize the interval of variability of the concentration $P(\underline{x})$. The different kriging estimates $i(\underline{x};z_k)^*$ are then pieced together to provide an estimate of the conditional cdf $F_{\underline{x}}(z|(N))$.

Note that if the kriging weights $\lambda_i(z)$ are data values-independent, the indicator estimates (11) are not; hence, the final estimate of the conditional cdf $F_x(z|(N))$ is data values-dependent. For each cut-off z_k, the kriging estimator (11) requires a different indicator covariance model:

$$C_I(h;z_k) = E\{I(\underline{x};z_k)\cdot I(\underline{x} + \underline{h};z_k)\} - F^2(z_k) \tag{12}$$

These various covariance models are inferred directly from the corresponding indicator data $i(\underline{x}_i;z_k)$, $i=1,\ldots,N$. The indicator kriging approach is said to be "non-parametric," in the sense that it draws solely from the data, not from any multivariate distribution hypothesis, as was the case for the multi-ϕ-normal approach.

These indicator covariance models allow differentiation of the spatial correlation of high-valued concentrations (cut-off z_k high) and low to background-valued concentrations (z_k low). In the particular case study underlying the Figure 3, it was found that high value concentration data were more spatially correlated (due to the plume of pollution) than lower value data.

Conclusions

Assessment of spatial distributions of pollutant concentrations is a very specific problem that requires more than blind mapping of these concentrations. Not only must the criterion of estimation be chosen carefully to allow zooming on the most critical values (the high concentrations), but also the evaluation of the potential error of estimation calls for a much more meaningful characteristic than the traditional estimation variance. Finally, the risks α and β of making wrong decisions on whether to clean or not must be assessed.

An answer to the previous problems is provided by the conditional distribution approach, whereby at each node $\underline{x}$ of a grid the whole likelihood function of the unknown value $p(\underline{x})$ is produced instead of a single estimated value $p^*(\underline{x})$. This likelihood function allows derivation of different estimates corresponding to different estimation criteria (loss functions), and provides data values-dependent confidence intervals. Also this likelihood function can be used to assess the risks α and β associated with the decisions to clean or not.

Literature Cited

1. Flatman, G. T. Proc. of Workshop on Environmental Sampling, EPA Las Vegas, February 1984.
2. Brooker, P. I. In "Geostatistics"; McGraw Hill, 1980; pp. 41-60.
3. Journel, A. G. In "Geostatistics for Natural Resources Characterization"; Verly G. et al., Ed., Reidel: Dordrecht, Netherlands, 1984; Part 1, pp. 261-270.
4. Anderson, T. W. "An Introduction to Multivariate Statistical Analysis"; Wiley & Sons: New York, 1958; p. 374.
5. Verly, G. W. Math Geology 1983, 15, 2, p. 263-290.
6. Parker, H. M.; Journel, A. G; and Dixon, W. L. Proc. of the 16th APCOM Symp., AIME, New York, 1979, pp. 133-148, ed.
7. Luenberger, D. L. "Optimization by Vector Space Methods"; Wiley & Sons: New York, 1969; p. 326.

8. Journel, A. G. Math. Geology 1983, 15, 3, p. 445-468.
9. Sullivan, J. In "Geostatistics for Natural Resources Charac-
    terization"; Verly, G., et al., Ed.; Reidel: Dordrecht, Nether-
    lands, 1984; Part 1, pp. 365-384.
10. Isaaks, E. H. MSc Thesis, Stanford University, 1984.

RECEIVED August 6, 1984

Detecting Elevated Contamination by Comparisons with Background

WALTER LIGGETT

National Bureau of Standards, Washington, DC 20234

The objective of the studies considered in this paper is detection of excess levels of a common soil contaminant. To meet this objective, these studies must include the sampling and measurement of a background region so that detection can be based on comparison of the background measurements with measurements on the suspected region. Thus, the background measurements are important in the studies considered, much more important than in other studies such as those aimed at determining the spatial extent of a highly contaminated area.

To compare the measurements from the two regions, the studies discussed use statistical methods based on the following model: The background measurements are observations drawn at random from some population. Since the background measurements consist of actual background concentrations perturbed by subsampling and measurement error, this model requires that both the actual background concentrations and the subsampling and measurement errors be random. The contamination in the suspected region is the sum of the background contamination and any excess contamination. Measurements on the suspected region are also perturbed by subsampling and measurement error. Thus, in the absence of excess contamination, the measurements from the suspected region are just further random observations from the background population. The two regions can be compared by asking whether the measurements from the suspected region could be observations from the same population as the background measurements. Detection follows from the conclusion that the two sets of measurements could not reasonably be from the same population. In statistical terms, detection is rejection of the null hypothesis that the two sets are from the same population.

Background measurements that are obtained from soil samples may have an asymmetrical distribution, that is, a skew distribution. Typically, the asymmetry is characterized by a positive skewness and by a higher probability of unusually large values than of unusually small values. This asymmetry may originate from the distribution of the contaminant over the background region or the procedure used to extract a subsample for laboratory analysis. Several authors discuss the relation of sampling and subsampling to the variability of measurements (1-4). These authors also review other work on sampling and subsampling.

Background measurements with a skew probability distribution generally require an appropriate statistical method. In particular, a statistical method based on appropriate assumptions about the upper tail of the background distribution is needed if the objective is to detect a few hot spots in the suspected region on the basis of the highest measurements from that region. In this context, the upper tail of the background distribution is important because detection should not occur when the highest measurements from the suspected region might reasonably be observations from the upper tail of the background distribution. Appropriate statistical methods can be based on transformation to normality (5,6). Other methods might be based on a skew distribution such as the gamma distribution (7).

In this paper, we discuss studies based on comparison with background measurements that may have a skew distribution. We discuss below the design of such a study. The design is intended to insure that the model for the comparison is valid and that the amount of skewness is minimized. Subsequently, we present a statistical method for the comparison of the background measurements with the largest of the measurements from the suspected region. This method, which is based on the use of power transformations to achieve normality, is original in that it takes into account estimation of the transformation from the data.

The Model and Its Validity

To provide a framework for the design of the study, let us specify the model for the measurements in statistical terms. The background measurements, which we denote by x_{Bi}, $i = 1,...,n_B$, are a random sample from some population. The measurements from the suspected region, denoted by x_{Si}, $i = 1,...,n_S$, are, in the case of no excess contamination, a second, independent random sample from this same population.

The study should be designed so that this model can be applied in the data analysis. This goal has various ramifications. First, the background region should be generally the same as the suspected region except that the background region must have no excess contamination. Second, the same sampling, subsampling, and measurement procedures must be used to obtain all the measurements. Third, these procedures should minimize skew in the background distribution and should reduce sensitivity of the measurements to any unavoidable extraneous differences between the two regions.

Consider first the choice of the background region. For concreteness, consider the case of large areas characterized by shallow pollutant deposition, one of the cases discussed by Mason (8). Clearly, the background region should have the same average contamination level as the suspected region would have were there no excess contamination. The background region should not differ from the suspected region in some characteristic such as soil type or vegetation that is extraneous to the question of excess contamination but might cause the two sets of measurements to differ. The contamination in the background region should have no large-scale spatial variations since two regions are not likely to be comparable if one exhibits large-scale variations. If the suspected region exhibits large-scale spatial variations, then it should be divided into more homogeneous subregions and a matching background region found for each subregion.

Ideally, the two regions should be identical in the absence of excess contamination. However, at best, the two regions will exhibit irregular spatial variations that are different. Under these circumstances, the points at which the samples are collected could be chosen in such a way that the comparison would be invalid. One way to guard against this possibility is random selection of the sampling points. If the regions differ only in their patterns of irregular spatial variation, then random selection could plausibly produce sets of observations that would appear to be from the same population. Although random selection from the background region seems reasonable, random selection from the suspected region might be rejected in favor of a sampling plan based on knowledge of how the excess contamination could have been deposited.

The design goal of insuring that the statistical model is appropriate obviously requires that the same physical procedures for sampling and subsampling be used in both regions. This design goal has other implications for the choice of the sampling and subsampling procedures. In particular, a sampling procedure that calls for the collection of a large amount of soil may result in measurements that are less sensitive to extraneous differences between the two regions (1). Another reason to choose large samples is to reduce the skewness of the background measurements. The size of the samples is limited by the size of the hot spots that are to be detected. In samples that are too large, the presence of the excess contamination might be washed out. Smith and James (1) discuss the biases inherent in the use of various sampling instruments. If these biases are the same for both regions, these biases are of less concern in a comparison than in other cases such as the estimation of an average.

Generally, only a subsample of each sample collected is chemically analyzed. To obtain these subsamples, we need a carefully designed subsampling procedure. In some cases, the difference in mass between the sample and the subsample is very large (3). For this reason, the subsampling procedure must include vigorous grinding and mixing so that the subsampling does not introduce too much variability and asymmetry. A review of various design criteria for the minimum subsample mass has been made (1). This mass is given in terms of the size to which the particles in the sample have been reduced and the amount of error that can be tolerated. These design criteria assume thorough mixing even though this is sometimes hard to achieve (9). The lack of thorough mixing may cause the subsampling procedure to be biased. In a comparison, this bias will balance out if it is the same for both regions. However, the bias might depend on some spurious factor that is not the same in both regions.

The choice of a chemical analysis procedure is also part of the design. We will not discuss the additional difficulties the chemical analysis procedure can introduce. In this paper, we assume that the chemical analysis procedure has adequate sensitivity and relatively small error.

The literature on sampling bulk materials is more specific and more quantitative than the general discussion presented above. Nevertheless, experiments on the sampling and subsampling procedures are often a necessary part of the design process. Liggett, et al. (10) discuss experiments on subsampling procedures. They consider one-to-twenty gram subsamples of plutonium-contaminated soil and show that despite vigorous grinding and mixing, the subsamples exhibit asymmetrical subsampling error. The question of what steps to take to reduce asymmetry in the distribution of the background measurements involves a cost-benefit tradeoff that can only be decided by experiments that show how much reduction is given by various modifications of the sampling and subsampling procedures.

A Method for Detecting Hot Spots

Various statistical methods can be used to compare the measurements from the suspected region with the background measurements. The power of these methods depends on how the excess contamination is distributed over the suspected region. Comparison of the means of the measurements from the two regions is appropriate if a uniform distribution of excess contamination is expected. Comparison of the maximum measurements from the suspected region with the background measurements is appropriate if a few hot spots with high contamination levels are expected. This latter case, which is the one considered in this section, is one in which positive skewness in the distribution of the background measurements cannot be ignored.

The best known approach to measurements with positive skewness is transformation. In environmental data analysis, the measurements are often transformed to their logarithms. In this paper, we consider power transformations with a shift, a set of transformations that includes the log transformation and no transformation at all (5). These transformations are given by

$$y = \begin{cases} [(x - \tau)^\lambda - 1]/\lambda & \text{for } \lambda > 0 \\ \\ \log (x - \tau) & \text{for } \lambda = 0. \end{cases} \tag{1}$$

This set of transformations depends on two parameters, the power parameter λ and the shift parameter τ. We restrict λ to non-negative values to avoid having a finite upper bound on the y values. (When λ is negative, the point $y = -1/\lambda$ corresponds to $x = \infty$.) Also, τ must be less than the smallest value of x. When τ is positive, we can think of the measurements as the sum of a constant component with level τ and a variable component. Since the constant component cannot have a negative level, restricting τ to positive values makes sense in some contexts. In soil sampling, we need a more general set of transformations than just the log transformation because the amount of skewness and therefore the proper transformation depends on the choice of sampling and subsampling procedure. The transformations given by Equation 1 seem to fulfill this need.

Although we usually do not know the values of λ and τ that transform the background measurements to normality, let us consider the case in which we do. Using Equation 1, we transform the measurements x_{Bi} and x_{Si} to y_{Bi} and y_{Si}, respectively. To compare the maximum measurement from the suspected region with the background measurements, we form the t statistics

$$t_j = (y_{Sj} - m_B)/(s_B (1 + 1/n_B)^{1/2}),$$ (2)

where m_B and s_B are given by

$$m_B = \sum_i y_{Bi}/n_B,$$

$$s_B = \left(\sum_i(y_{Bi} - m_B)^2/ (n_B-1)\right)^{1/2},$$ (3)

and perform the test using $\max_j t_j$ as the test statistic. To perform the test, we need a critical value c with which to compare the test statistic. As in the usual statistical hypothesis testing, c is chosen so that the probability of the test statistic exceeding c when there is no excess contamination is α. This is the probability of false detection. Although an exact value of c can be obtained by numerical integration, an approximate value that is just a little too high can be obtained more simply. This value is given by the $100(1 - \alpha/n_S)$ percentile of the t distribution with $n_B - 1$ degrees of freedom (11). Denoting the cumulative t distribution function by $F_t(t; n_B-1)$, we have

$$F_t(c; n_B-1) = 1 - \alpha/n_S.$$ (4)

Since we do not know the proper values for λ and τ, we need a way of judging plausible values of λ and τ from the data. We do this by testing the transformed background measurements for normality. Our choice of a test for normality is the probability plot correlation coefficient r (12). The coefficient r is the correlation between the ordered measurements and predicted values for an ordered set of normal random observations. We denote the ordered background measurements by $y_{B(i)}$, where $y_{B(1)} \leqslant y_{B(2)} \leqslant \ldots \leqslant y_{B(n_B)}$. We denote the predicted values with which the ordered measurements are correlated by M_i. The computation of these values is discussed by Filliben (12). A conceptual definition of the M_i follows from consideration of a set of numbers drawn at random from the standard normal distribution, the one with mean zero and standard deviation one. Ordering this set of numbers gives a sequence called order statistics. The M_i are the medians of the distributions of these order statistics. Since M_i predicts the value of the ith order statistic from the standard normal distribution, $\mu + \sigma M_i$ predicts the value of the ith order statistic from the normal distribution with mean μ and standard deviation σ. Thus, if the ordered transformed measurements are from a normal distribution and they are plotted versus the M_i, they should fall near a straight line. How close they come to a straight line is measured by the probability plot correlation coefficient

$$r = \frac{\sum_i M_i(y_{B(i)}-m_B)}{\left[\left(\sum_i M_i^2\right)\left(\sum_i(y_{B(i)}-m_B)^2\right)\right]^{1/2}}.$$ (5)

Filliben (12) tabulates $F_r(r)$, the distribution of r.

   To construct a test for general n_S and (λ, τ) unknown, we start with a test for $n_S = 1$ that simultaneously tests values of x_{S1}, λ, and τ. From this simultaneous test, we derive a test for unknown (λ, τ) and $n_S = 1$. Finally, we generalized to arbitrary values of n_S.

   We begin with a simultaneous test of the null hypothesis that the single measurement x_{Sj} is drawn from the background population and that λ and τ provide the proper transformation to normality. Our simultaneous test, which is based on t_j and r, rejects the null hypothesis if

$$t_j > c(r) \qquad\qquad (6)$$

where

$$c(r) = \begin{cases} k\;[(1 - .97r)^{-1/2} - 1] & \text{for } r \geqslant r_{.005} \\[2ex] c(r_{.005})\;[1 - 2\;(r_{.005}-r)/(r_{.005}-r_{min})] & \text{for } r < r_{.005}, \end{cases} \qquad (7)$$

$F_r\;(r_{.005}) = 0.005$ and r_{min} is the minimum value of r (12). The function $c(r)$ was chosen through Monte Carlo experiments. The value of k is chosen so that the probability of $t_j > c(r)$, when x_{Sj} is drawn from the same population as the background measurements, is α. This probability is given by $1 - \int_0^1 F_t(c(r); n_B-1)\;dF_r(r)$.

   The simultaneous test given by Equations 6 and 7 leads to a test appropriate for (λ, τ) unknown. The (λ, τ)-unknown test rejects the null hypothesis that x_{Sj} belongs to the background population if $t_j > c(r)$ for all (λ, τ). Since this test rejects the null hypothesis only if Equation 6 is satisfied for the true value of (λ, τ), this test has no greater probability of false detection than the simultaneous test. Thus, the (λ, τ)-unknown test is conservative in the sense that the probability of a false detection is less than α if the probability of false detection for the simultaneous test is α.

   To test all the measurements from the suspected region, we choose the value of k in Equation 7 that satisfies

$$\int_0^1 F_t(c(r); n_B -1)\;dF_r(r) = 1 - \alpha/n_S. \qquad (8)$$

and proceed as above. We reject the null hypothesis that all the x_{Sj} are from the background population if $\max_j t_j > c(r)$ for all (λ, τ). This generalization from $n_S = 1$ to arbitrary n_S is based on the same principle as the use of Equation 4 in the (λ, τ) known case. This generalization is conservative in the sense that the actual probability of false detection is less than or equal to α.

   This test requires two computations. First, we must determine k, the parameter in $c(r)$. To do this, we must choose a numerical method for evaluating Equation 8. If α/n_S is not too small, we can use the tabulation of $F_r(r)$ provided by Filliben (12). We approximate Equation 8 by

$$0.0025 + \sum 0.5\;[F_t(c(r_{i+1})) + F_t(c(r_i))]$$
$$\cdot\;[F_r(r_{i+1}) - F_r(r_i)] = 1-\alpha/n_S \qquad (9)$$

where the r_i are the points at which Filliben ($\underline{12}$) evaluates F_r. To solve Equation 9 for k, we need an algorithm for evaluating F_t ($\underline{13}$) and an algorithm for finding roots. Since the left side of Equation 9 is an increasing function of k, a simple root-finding algorithm such as repeated halving of an interval known to contain k is adequate. Since Filliben ($\underline{12}$) only tabulates $F_r(r)$ down to 0.005, very small values of α/n_S require a Monte Carlo evaluation of Equation 8. The second computation is the test itself. This can be arranged as the computation of a critical value x_I with which $\max_j x_{Sj}$ is compared. The critical value x_I is $\max_{\lambda,\tau} x_u$, where

$$x_u = \begin{cases} \tau + (1 + \lambda y_u)^{1/\lambda} & \text{for } \lambda > 0 \\ \\ \tau + \exp(y_u) & \text{for } \lambda = 0 \end{cases} \qquad (10)$$

and

$$y_u = m_B + c(r)(1 + 1/n_B)^{1/2} s_B. \qquad (11)$$

To perform the maximization over (λ, τ), we need an algorithm such as the Nelder-Mead simplex search ($\underline{14}$). An alternative that is adequate in many cases is a simple search over a (λ, τ) grid. The critical value x_I has an interpretation of its own. It is the upper bound on a simultaneous prediction interval for n_S as yet unobserved observations from the background population.

Evaluation of the Method

The previous section presents a test that is based on the assumption that for some (λ, τ) Equation 1 transforms the background measurements to nomality but is otherwise conservative. The test contains no explicit restriction on n_B or n_S except $n_B > 2$. However, the test cannot be expected to be satisfactory for all values of n_B and n_S. First, the test is based on extrapolation of the distribution of the background measurements to higher values than are represented in the data. If extrapolation is carried too far, the results will not be satisfactory. Second, the test is conservative and may be too conservative for some values of n_B and n_S. In this section, we present an example based on analogous data that shows values of n_B and n_S for which the test is useful.

Before looking at the example, consider a very simple alternative to the test just described. This alternative is valid regardless of what the distribution of the background measurements is. The alternative is to declare a detection if the largest of the measurements from the suspected region is larger than the largest of the background measurements. The probability of false detection with this procedure is $n_S/(n_B + n_S)$ ($\underline{15}$). This formula shows that n_B must be much larger than n_S if the probability of false detection is to be suitably small.

The data considered are blank measurements made as part of a study of trace quantities of heavy metals dissolved in the water of the Chesapeake Bay ($\underline{16}$). While obtaining and processing the Bay samples, samples of high-purity sub-boiling distilled water were handled in the same way. These blank samples were exposed to the same sources of processing contamination as the Bay samples. Thus, the question of whether the measurements on the Bay samples are due to no more than the processing contamination might be raised. This question is analogous to the detection of excess soil contamination. (Fortunately, the measurements on the Bay samples were well above the measurements on the blank samples.) We consider three metals, Cobalt, Iron, and Scandium. The twenty-four blank measurements for each of these metals are given in Table I.

In our application of the transformation given in Equation 1 to these data, we restrict τ to positive values. This restriction is based on models of the various sources of processing contamination. Possible sources of contamination include the chemical reagents which might add a constant level to the blank and air borne particles which might add a variable level with positive skewness. There does not seem to be any reason to include a constant level that is negative. Therefore, we have adopted this restriction.

Each of these data sets is skewed, yet each can be transformed to normality. With no transformation applied, the probability plot correlation coefficients for the Co, Fe, and Sc data sets are 0.855, 0.857, and 0.987, respectively. For Co and Fe, the hypothesis of normality is rejected at the 0.5 percent level (12). On the other hand, the maximum probability plot correlation coefficients are 0.993, 0.990, and 0.993 for Co, Fe, and Sc, respectively. The maxima occur at (λ,τ) = (0,0.0048), (0,0.42), and (0.457,0), respectively. These maxima are so high that they provide no evidence that the range of transformations is inadequate. Note that the (λ,τ) values at which the maxima occur correspond to log transformations with a shift for the Co and Fe and nearly a square-root transformation for the Sc.

Pretend these data sets are the background measurements, and let n_S = 5. If we use the distribution-free test and compare the maximum measurement from the suspected region with the maximum measurement from the background region, we will have a probability of false detection of 5/29 (= 0.17). This probability may be too high. To obtain a probability of false detection of 0.05, we need the test discussed in the previous section. From Equation 9, we find that k = 0.7753. From Equations 10 and 11, we obtain x_I = 0.228, 6.10, and 0.00033 for Co, Fe, and Sc, respectively. These values are substantially higher than the largest of the background measurements. This is due in part to the conservativeness of the test. Monte Carlo experiments on the test discussed in the previous section suggest that the k value 0.7753 corresponds to α/n_S = 0.005 instead of 0.010. Thus, these values of x_I correspond to n_S = 10 instead of 5.

The test presented in the previous section is useful when a smaller probability of false detection is needed than is provided by the distribution-free test. However, the test in the previous section is no panacea. Reduction of the skewness through proper choice of sampling and subsampling procedures is an alternative that may have much more potential for improving the study.

Table I.   Blank Concentrations for Dissolved Cobalt, Iron, Scandium
(ng/mL) Given as Stem-and-Leaf Displays (Tukey 1977) (6)

Cobalt

.00**	53,53,58,58,63,66,72,73,75,78,81,87,95,95
.01	00,10,10,23,47,60
.02	40,80
.03	00
.04	00

Iron		Scandium	
0.**	59,68,75,79,80,89,90,95	.0000*	5,7,7,8,8,9,9
1.	00,02,10,10,10,20,24,25,30	.0001	0,0,0,1,1,2,2,3,4,4,4
1.	50,50,53,70,88	.0001	5,5,6,6
2.	03	.0002	0,0
2.			
3.			
3.	70		

Note that the actual values of the measurements can be obtained by
combining the entries on the two sides of the vertical line.   For
example, the first Cobalt measurement is .0053 ng/mL.

Literature Cited

1. Smith, R.; James, G. V. "The Sampling of Bulk Materials"; The Royal Society of Chemistry: London, 1981.
2. Kratochvil, B.; Taylor, J. K. Analytical Chemistry 1981, 53, 924A-938A.
3. Ingamells, C. O. Talanta 1974, 21, 141-155.
4. Ingamells, C. O.; Switzer, P. Talanta 1973, 20, 547-568.
5. Box, G. E. P. and Cox, D. R. Journal of the Royal Statistical Society, 1964, Series B, 26, 211-252.
6. Tukey, J. W. "Exploratory Data Analysis"; Addison-Wesley Publishing Company: Reading, Mass., 1977.
7. Bain, L. J. "Statistical Analysis of Reliability and Life-Testing Models"; Marcel Dekker: New York, 1978.
8. Mason, B. J. "Preparation of Soil Sampling Protocol: Techniques and Strategies," Report EPA-60014-83-020, U.S. Environmental Protection Agency, 1983.
9. Williams, J. C. Powder Technology 1976, 15, 245-251.
10. Liggett, W. S.; Inn, K. G. W.; Hutchinson, J. M. R. Environmental International 1984, to appear.
11. Hahn, G. J. Journal of the American Statistical Association 1970, 65, 1668-1676.
12. Filliben, J. J. Technometrics 1975, 17, 111-117.
13. Kennedy, W. J.; Gentle, J. E. "Statistical Computing"; Marcel Dekker: New York, 1980.
14. Nash, J. C. "Compact Numerical Methods for Computers: Linear Algebra and Function Minimization"; John Willey and Sons, New York, 1979.
15. Hahn, G. J.; Nelson, W. Journal of Quality Technology 1973, 5, 178-188.
16. Kingston, H. M.; Greenberg, R. R.; Beary, E. S.; Hardas, B. R.; Moody, J. R.; Rains, T. C.; Liggett, W. S. "The Characterization of the Chesapeake Bay: A Systematic Analysis of Toxic Trace Elements," Report NBSIR 83-2698, National Bureau of Standards, 1983.

RECEIVED August 14, 1984

Author Index

Subject Index

Production and indexing by Karen McCeney
Jacket design by Pamela Lewis

Elements typeset by Hot Type Ltd., Washington, D.C.
Printed and bound by Maple Press Co., York, Pa.

RECENT ACS BOOKS

"Computers in the Laboratory"
Edited by Joseph Liscouski
ACS SYMPOSIUM SERIES 265; 136 pp.; ISBN 0-8412-0867-0

"The Chemistry of Low-Rank Coals"
Edited by Harold H. Schobert
ACS SYMPOSIUM SERIES 264; 328 pp.; ISBN 0-8412-0866-2

"Resonances in Electron-Molecule Scattering, van der
Waals Complexes, and Reactive Chemical Dynamics"
Edited by Donald G. Truhlar
ACS SYMPOSIUM SERIES 263; 536 pp.; ISBN 0-8412-0865-4

"Seafood Toxins"
Edited by Edward P. Ragelis
ACS SYMPOSIUM SERIES 262; 473 pp.; ISBN 0-8412-0863-8

"Computers in Flavor and Fragrance Research"
Edited by Craig B. Warren and John P. Walradt
ACS SYMPOSIUM SERIES 261; 157 pp.; ISBN 0-8412-0861-1

"Polymers for Fibers and Elastomers"
Edited by Jett C. Arthur, Jr.
ACS SYMPOSIUM SERIES 260; 448 pp.; ISBN 0-8412-0859-X

"Treatment and Disposal of Pesticide Wastes"
Edited by James N. Seiber and Raymond F. Krueger
ACS SYMPOSIUM SERIES 259; 384 pp.; ISBN 0-8412-0858-1

"Stable Isotopes in Nutrition"
Edited by Judith R. Turnlund and Phyllis E. Johnson
ACS SYMPOSIUM SERIES 258; 240 pp.; ISBN 0-8412-0855-7

"Bioregulators: Chemistry and Uses"
Edited by Robert L. Ory and Falk R. Rittig
ACS SYMPOSIUM SERIES 257; 286 pp.; ISBN 0-8412-0853-0

"Polymeric Materials and Artificial Organs"
Edited by Charles G. Gebelein
ACS SYMPOSIUM SERIES 256; 208 pp.; ISBN 0-8412-0854-9

"Pesticide Synthesis Through Rational Approaches"
Edited by Philip S. Magee, Gustave K. Kohn, and Julius J. Menn
ACS SYMPOSIUM SERIES 255; 351 pp.; ISBN 0-8412-0852-2

"The Chemistry of Solid Wood"
Edited by Roger M. Rowell
ADVANCES IN CHEMISTRY SERIES 207; 588 pp.; ISBN 0-8412-0796-8

"Polymer Blends and Composites in Multiphase Systems"
Edited by C. D. Han
ADVANCES IN CHEMISTRY SERIES 206; 400 pp.; ISBN 0-8412-0783-6